Life in the Sea

MARTY SNYDERMAN

CONSULTANT
M. TUNDI AGARDY, PH.D.

PUBLICATIONS INTERNATIONAL, LTD.

Marty Snyderman is a world-renowned marine photographer and cinematographer. His work has been featured by the Columbia Broadcasting System, the British Broadcasting Corporation, the Audubon Society, the National Geographic Society, *Nova, Newsweek,* and the National Wildlife Federation. His writing and photography have appeared in a number of publications including *Ocean Realm* and *Sea Frontiers.*

M. Tundi Agardy, Ph.D., is Senior Conservation Scientist at the World Wildlife Fund, with a special emphasis on the protection of coastal ecosystems and marine species worldwide. She also serves as Vice-Chairperson of the Commission of Ecology for the World Conservation Union. She has done work for the Woods Hole Oceanographic Institution and the British Broadcasting Corporation.

Cover: A pair of clownfish in an anemone. *Opposite page:* A trigger-fish.

Photo Credits:

Front cover: **Soames Summerhays/Biofotos.**

Animals Animals: Doug Allan: 34, 220 (top); Doug Allan/Oxford Scientific Films: 32, 208; G.I. Bernard/Oxford Scientific Films: 39, 148 (bottom); W. Gregory Brown: 16, 66, 102; M.A. Chappell: 214; Ken Cole: 199, 205; Tony Crabtree/Oxford Scientific Films: 172; E.R. Degginger: 8, 87; David C. Fritts: 190; Mickey Gibson: 6, 186; Johnny Johnson: 196, 210, 196-197; R. Kolar: 238-239; Gerald L. Kooyman: 33; Alison Kuiter/Oxford Scientific Films: 119; Rudie Kuiter/Oxford Scientific Films: 189; Gerald Lacz: 193; Zig Leszczynski: 77, 118; Tony Martin/Oxford Scientific Films: 225; Brian Milne: 213; Ralph A. Reinhold: 203; Carl Roessler: 137, 174; Bruce Watkins: 115, 155 (bottom); James D. Watt: 28-29, 167, 180, 238, 240; Fred Whitehead: 156; Capt. Clay H. Wiseman: 71, 91, 126, 185; **Heather Angel Photography:** 64, 74, 97, 202-203, 211, 223, 234-235; **Warren D. Barrett:** 227; **Biofotos:** J.W.H. Conroy: 209; Soames Summerhays: 114 (left), 152, 153, 206; Ian Took: 125; **Stephen Myers/FPG International:** 176; **Marty Snyderman:** 7, 13, 21, 44, 45 (top), 46, 57, 62, 70, 81, 84, 89, 93, 94, 104, 105, 113, 121 (top), 122, 123, 129, 130, 131, 132, 133, 138, 140, 144, 150, 151, 154 (bottom), 158, 159, 160, 161, 162, 163, 170, 171, 173, 182, 183, 187, 192, 195, 207, 228, 229, 230, 231, 236, 237; **Tom Stack & Associates:** Mike Bacon: 4, 61, 110, 112; D. Holden Bailey: 117, 216; W. Perry Conway: 31; Gerald & Buff Corsi: 58; Gerald Corsi: 108, Dave B. Fleetham: 48 (left & right), 49 (bottom), 50, 60, 85, 103, 116, 146, 164, 166, 217; Jeff Foott: 201, 215, 219; Garoutte: 106; Thomas Kitchin: 9, 222-223, 224 (center); Larry Lipsky: 12, 82, 139; Joe McDonald: 109; Gary Milburn: 100; Randy Morse: 18, 22, 26, 27, 28, 55, 155 (top), 168, 169, 188; Mark Newman: 212, 218; Brian Parker: 48 (bottom), 49 (left), 53, 65, 98, 99, 221 (top); Tammy Peluso: 10, 134; Jack Reid: 124, 149; Ed Robinson: 69 (top), 96, 111, 165, 175, 179, 181, 184, 194; Kevin Schafer: 198; Mike Severns: 11, 88, 141, 145; John Shaw: 200; Larry Tackett: 49 (top), 128; Denise Tackett: 157; Dave Watts: 30; Robert Winslow: 204; **Richard Todd:** Table of contents, 47, 178; **The WaterHouse Photography:** Stephen Frink: 75; Neil McDaniel: 148 (top) **Norbert Wu:** 5, 14, 15, 17, 19, 20, 23, 35, 36, 37 (top), 38, 40, 41, 42, 43, 45 (right & bottom), 51, 52, 54, 56, 58-59, 63, 67, 68, 69 (bottom), 72, 73, 76 (bottom), 78, 79, 80, 83, 86, 90, 92, 95, 101, 107, 114, 121 (bottom), 127, 135, 136, 142, 143, 147, 154 (top), 156 (bottom), 177, 191, 220 (bottom), 221 (bottom), 226, 232, 233; Bob Cranston: 24, 25, 76 (top), 120.

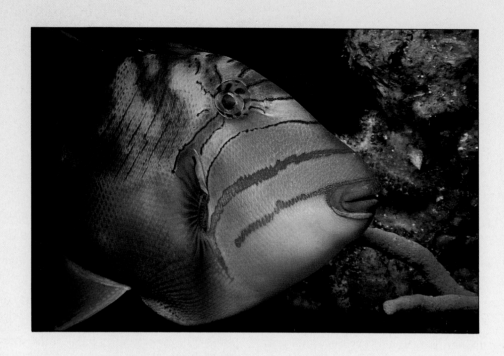

Introduction

THE OPAQUE OCEANS HARBOR A hidden world of wonder. Nature's most prolific studio has produced a vast array of species and a complex network of interactions, making the oceans the most diverse and productive areas on the planet. Water—elemental and crucial to life—has provided countless opportunities for experiments in evolution that have no parallel on land, and this process has been going on in the oceans since life on Earth first began hundreds of millions of years ago. The incredible diversity of sea life is underscored by the distribution of phyla (the major categories distinguishing living things): 32 of 33 known phyla are found in the oceans, and 15 of them live exclusively in the marine world.

The myriad organisms in the sea attest to the numerous ways of living that the ocean environment permits. We tend to think of the seas as being monochromatic and homogenous: the big blue, a vast but uniform and unchanging place. Actually, the oceans and coasts offer a startling array of habitats that change continually, presenting marine creatures with ever new options for survival. Rocky and sandy shorelines, mud flats and seagrass beds, kelp forests and coral reefs, the

mid-water environment of the open ocean and the benthos of the deep sea—these habitats are as different from one another as forests are from deserts.

This complex marine realm presents its occupants with both challenges and opportunities. The challenges have to do with being in the right place at the right time and with weathering the sometimes extreme environmental conditions, such as the dim, deadly cold of the deep sea or the wave-pounded temperate rocky shorelines. The opportunities arise because water is an ideal medium for life: the density of seawater eases the constraints imposed by gravity, resulting in some truly unusual body shapes.

Marine organisms range in complexity from bacteria that live on the sea surface microlayer, to minute one-celled plants collectively known as phytoplankton, to small animals (zooplankton) found grazing the plants in the water column, to the more familiar macroalgae and seaweeds, sponges, jellyfish, corals and anemones, mollusks, crustaceans, fish, marine reptiles like sea snakes and sea turtles, and finally seals, manatees, whales, and other marine mammals. These organisms are interconnected across vast ocean distances, so that entire seas pulsate with indivisible webs of life stretching through them.

Opposite page: *Damselfish, a striped orange clownfish, and a huge anemone in the Red Sea.* Above: *In just a few square feet, a typical coral reef might contain dozens of species.*

The connections in this web of life are sometimes subtle. Living coral animals lay down reef skeletons that provide a habitat for invertebrates and fish, while kelp forests provide a similar framework for cold-water organisms. Many marine animals are highly dependent on each other: clownfish reside in the safety of the stinging tentacles of sea anemones and provide them with food through their own leftovers; cleaner fish make meals of the parasites that trouble larger fish, and for performing this favor, the larger fish resist swallowing the vulnerable cleaners. Such behavior links seemingly unrelated organisms and underscores the delicate balance that has evolved in marine communities.

The oceans are vast, almost bigger than our comprehension allows, and certainly they are still full of unexplored mysteries. Over many generations, humans have slowly learned to reap the sea's bounty and to take advantage of the resources and space it offers. Using our ever-advancing technologies, we seem able to explore and harvest the seas efficiently, but not always wisely. Each time we deplete resources or degrade the sensitive habitats that serve as home to sea life, we risk undermining a system that has served us well. As individuals we may respect the oceans and regard them with awe, but collectively we sometimes set aside these views in order to exploit the Earth's greatest storehouse. We may now be at a crossroads with respect to the future of the oceans. While signs of environmental degradation in the seas are rampant, public awareness of the beauty and bounty of the oceans continues to grow. As people come to appreciate the critical functions that marine systems serve in maintaining the Earth, the chances for a future in which oceans are conserved for the benefit of all increase greatly.

Such an appreciation of the oceans truly comes when you see marine animals and plants with your own eyes, smell and taste the salty brine, witness firsthand how the inky ocean fluid plays with the light of the sun, and take in the enormous scales and vastness of a variety of seascapes. The following pages will serve to introduce you to life in the seas and perhaps reinforce your sense of wonder at the spectacular array of marine creatures and habitats. You'll venture to five main ocean areas: the colorful coral reefs, the productive and complex temperate seas, the blue world of the open ocean, the frigid polar seas, and the murky deep seabed. You'll then be introduced to a diversity of life-forms, ranging from sponges, corals, worms, and mollusks to marine reptiles, fish, and mammals. It is our hope that this book whets your appetite and encourages you to take a closer look at ocean life, and in so doing increases your appreciation for the sea and its impressive inhabitants.

M. Tundi Agardy

M. Tundi Agardy
Senior Scientist
World Wildlife Fund

Places in the Sea

THE WORLD'S OCEANS ARE home to more than 200,000 known species of plants and animals, and they are all interrelated in some way. Any one species is dependent on a variety of others for food, for protection, sometimes even for reproduction. These creatures converge in a variety of marine environments to play out the drama of their lives, and each has evolved a unique form and distinctive survival strategy that allow it to win a place in its habitat.

TROPICAL REEFS

Very few natural settings are as stunning as the oceans' coral reefs. The massive coral structures, the dazzling colors, the subtle beauty of the reef creatures, and the diverse life within the coral kingdom are truly overwhelming. Excepting tropical rain forests, coral reefs are the most diverse habitats on Earth, sheltering a vast number of marine species.

Coral reef communities occur primarily in two sectors of the world: the tropical waters of the western Atlantic Ocean, which includes the Caribbean Sea, and the tropical Indo-Pacific, a region that stretches from the east coast of Africa to Australia and the central Pacific Ocean. Other coral reefs have taken hold near Hawaii, in

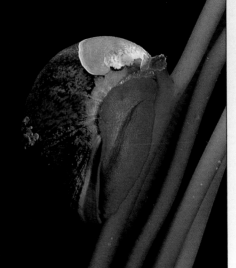

Mexico's Sea of Cortez, along Central America's west coast, and in the Red Sea.

Unlike any other habitat on Earth, coral reefs form over countless millennia through the efforts of millions and millions of tiny animals. Corals are dynamic, living organisms. Every hour, every day, some coral polyps perish and others spring to life. In a slow but steady process, coral reefs grow and alter their shape over countless generations to create tropical habitats brimming with life.

TEMPERATE SEAS

The temperate seas are vast expanses of ocean lying between the tropical and polar regions. The water surface temperature there avoids the extremes of some other marine habitats, ranging from the low 40s to the low 70s. Reef-building corals cannot survive in the cool waters, so rocky areas and kelp forests serve as the primary havens for marine life.

Kelp is a form of brown algae that most people only see once it has torn loose from the seafloor and washed ashore to decay. Growing in the ocean, though, a healthy kelp forest, or bed, rivals the beauty and wildlife of

🐚 Opposite page: *The broad range of species seen here is typical of most reef communities.* This page: *This kelp snail is one of the many creatures common in temperate waters.*

mature forests on land. A healthy forest of giant kelp can be home to more than 800 species of marine animals. A single plant can support more than a million organisms. Many are microscopic, but the sheer numbers show just how prolific a healthy forest can be.

The large base structures in the temperate zones consist of rock, mud, and sand. Rocky areas offer their inhabitants a surface for attachment and caves and crevices for protection. They also serve as wave breaks, providing respite from the violent, churning open sea. The rocks create a backdrop for lobsters, snails, scallops, eels, sharks, and a huge variety of fish.

OPEN WATERS

The open ocean offers no solid surfaces that life can cling to. It lacks any obvious sanctuaries or refuge from its uniformity. It has only water—vast, ever moving, and everywhere the same. There is no place to rest and no place to hide. Still, hundreds of species thrive here, endlessly cruising the ocean's surface waters. Home is wherever they happen to be.

In many areas of the open sea, life may be abundant one day and absent the next. The creatures that live here are transient or migratory, always on the move. Scientists cannot predict where any animals will be or when they will be there. Over time, though, people have begun to develop an accurate picture of life in the open ocean.

As in most marine settings, the open-ocean food chains rest on a base of plankton, microscopic plants and animals that drift through the seas. Large concentrations range from only a few square yards to a few hundred square miles. Some zooplankton spend their entire lives traveling with the currents in the open ocean. Others are the larvae of fish, lobsters, and other animals that move to different habitats as adults. Flourishing plankton serves as prey to many species,

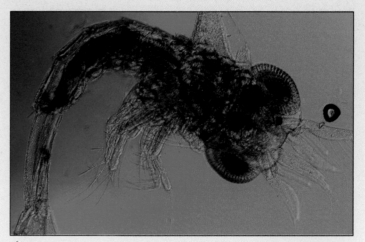

Above: *This amphipod is one of the countless planktonic species that form the base of most marine food chains.*

including herring, anchovies, sardines, shad, and other fish. These creatures in turn attract larger predators—bonito, yellowtail, barracuda, albacore, and jacks. And the mid-sized predators draw still larger hunters, such as swordfish, sailfish, marlin, dolphins, sharks, and toothed whales. All these creatures constantly travel the great tracts of open ocean, tirelessly seeking food and fleeing their predators. They are truly the nomads of the sea.

THE POLES

Few places on Earth are more forbidding than the polar regions. On the warm days, the temperatures approach freezing. The landscape seems carved of ice and snow. Some of the fiercest storms on Earth ravage this land, with howling, frozen gales blowing in all directions at once. Bleak, inhospitable, and frigid, the Arctic and Antarctic circles seem designed to keep life from taking hold.

Despite these conditions, polar seas are some of the most life-filled ocean regions on Earth. Much of the life is plankton, countless millions of tons of microscopic plants and animals that drift with the currents. But these organisms support a network of higher animals that have conquered their harsh surroundings.

🐚 Above: *A killer whale splashes in the waters off the coast of British Columbia, Canada.*

Crabs, lobsters, sea stars, barnacles, squid, and countless other invertebrates populate the polar seas. Huge schools of fish roam the Arctic and Antarctic waters, feeding and being fed upon. Marine mammals, too, are important members of the polar community. Seals, sea lions, walruses, polar bears, and whales dominate the regions.

The harsh conditions make polar life difficult to study. Humans have only glimpsed the beauty of this frozen world. The animals that live here, though, have faced the harshest of conditions and learned to persevere.

THE DEPTHS

Thousands of feet below the surface, the waters of the ocean are cold, still, and dark. Almost no light can penetrate at such depths, and the water temperature hovers near freezing. Still, life takes hold even here.

The harsh conditions of the abyss have generated some of the most nightmarish, bizarre-looking creatures on Earth. Many are scavengers that subsist on the remains of other animals. Some swim up to the mid-water regions to hunt and then return to the safety of the deep. Others are brutal predators that stalk the depths and consume most any living thing they encounter.

Opposite page: *Fairy basslets flit across a coral reef off Fiji. Hard corals grow in large branching or mounding colonies to form the basic structure of the reef. Most reef communities support a mixture of coral species that often grow quite near one another. Competition for food, space, and sunlight can be intense, and active battles occur between the competing corals. Over time, the resulting coral growth patterns form the reef's different zones—the seaward wall, the fore reef, the reef crest, the back reef, the rubble zone, and the sand patches. Different species of marine life settle in the zones that provide the living conditions they specifically need.*

Right: *A branching pink sea fan, a yellow coral colony, several brightly colored fishes, and a spidery black crinoid (upper right) are just a few of the animals that make their home in a tropical coral reef. Surely the most prominent vertebrates in the reef community are the many species of fish—eels, sharks, rays, butterfly fish, groupers, triggerfish, gobies, and sea horses. Reptiles such as marine snakes and sea turtles also live in the reef waters. Magnificently colored sea whips, crabs, sea stars, worms, sea urchins, snails, oysters, and anemones are among the more striking invertebrates of the reef. Less obvious, perhaps, but equally important are reclusive animals such as shrimp, lobsters, octopi, and squid.*

🐚 Above: *An elegantly patterned queen angelfish seeks refuge in a reef crevice. The tall thin body of the angelfish allows it access to narrow openings in the reef structure that many stout-bodied predators cannot penetrate. The clefts and fissures of the reef are essential to the survival of a great many species. They serve as home, hunting grounds, breeding territory, and sanctuary to a variety of animals. Some creatures wander across the reef, taking refuge in any available opening when the need arises. Others stake out a definite territory and defend it aggressively against all comers.*

🐚 Opposite page: *A pattern of fine spots helps this moray eel blend into its mottled environment as it makes a rare appearance outside its cave. The sinuous morays use the reef's many nooks and crannies to hide as they lie in wait for unsuspecting prey. They rely on surprise and a quick strike to capture their favorite foods—fish, octopi, and lobster. Their well-developed sense of smell alerts them when potential meals wander too close to the openings of their shadowy lairs.*

🐚 Opposite page: *Invertebrates, such as this fanworm perched on a coral colony, are among the many reef creatures that are active at night. Shrimp, hermit crabs, and squid openly roam the reef to hunt and forage in the evening. A vital part of the complex coral reef community is the daily cycle that sees one group of feeders retire to safety while another group begins to roam the reef. These transition periods around sunrise and sunset are especially active times. Jacks, barracuda, and other large predators actively hunt in these low-light conditions.*

🐚 Above: *The soldierfish is another of the many fish species that hunt and forage under cover of darkness. The soldierfish's large eyes give it excellent night vision. The daily feeding cycle of a reef enables the inhabitants to best use the available space. Except for areas claimed by territorial creatures, a reef's nooks and crannies house both daytime and nighttime occupants. Shortly after a nocturnal soldierfish leaves a protective reef crevice, a day-feeding angelfish might slip into the same hole to rest safely for the night.*

🐚 Opposite page: *A garden eel rises snakelike from its burrow in the sandy sea bottom to face into the current and feed. Though at first glance the sandy flats and rubble zones found next to coral reefs seem barren, they actually provide an important habitat for a whole community of organisms. Hogfish and other species excavate the seafloor, digging and blowing sand away as they scrounge for small mollusks and crustaceans. Sand tilefish, gobies, blennies, razorfish, stingrays, and some sea bass also reside in the sand and rubble.*

🐚 Above: *A sandy patch in shallow tropical waters provides the location for a large gathering of sand dollars. Members of the urchin family, sand dollars partially bury themselves in the seafloor and filter planktonic food from the gentle offshore currents that wash over them. A sand dollar's body is covered with a multitude of small flexible spines that draw food particles from the nutrient-rich waters and pass them toward the centrally located mouth.*

Left: *A lone garibaldi warily makes its way through a kelp forest off San Clemente Island in California. Like living, moving pillars in the sea, giant kelp fronds sway rhythmically with the passing swells, reaching up from the seafloor to the surface and the life-giving sun. The spaces between the kelp teem with fish and other marine life, and the plants themselves house a great many more creatures. Kelp plants typically grow near one another. Individual plants of some species can grow to 200 feet, and together they form dense undersea forests that can cover up to 10 square miles.*

Opposite page: *Sea otters abound in kelp forests, where they can find rich supplies of their favorite foods, including sea urchins. Frequently, they bring their prey to the surface and lie on their backs amid the floating kelp fronds as they dine. The spiny sea urchins can wreak havoc on a kelp forest by eating through the plants at their base. No longer anchored, the plants drift free in the water, snagging and ripping up other plants as they go. The otters, with their huge appetites and their taste for urchins, help to keep the urchin population in check and thereby maintain healthier kelp forests.*

🐚 Above: *This tiny octopus lives within a gnarled kelp holdfast. The holdfast is the bottom portion of the plant consisting of sturdy strands called haptera that grip rocky surfaces to secure the plant. The holdfast habitat contains many microscopic organisms as well as crustaceans, brittle stars, hydroids, bryozoans, and sea cucumbers. Above the holdfast, the stemlike stipes of the plant rise up through the water. Kelp bass, rockfish, clingfish, and numerous invertebrates live in this midwater habitat. At the surface, the plants' leafy fronds spread out across the water and gather sunlight. Fish, crabs, mollusks, and other invertebrates find a home in the lush surface canopy.*

🐚 Opposite page: *This kelp snail and the barnacle attached to its shell live in the mid-water habitat of the kelp forest. Snails sometimes gather by the hundreds in the middle regions, feasting on rotting or dead pieces of the kelp plant. Visible in the right portion of the photo is one of many gas-filled bulbs that help to keep the plant afloat. Together, the bulbs buoy the kelp fronds, suspending them in mid-water and allowing the plant to reach the surface and the sunlight that is so vital to its survival.*

🐚 Opposite page: *Competition within the rocky areas of temperate seas is intense. Predatory sea stars, such as these giant spined and ochre stars found off the Mexican coast, pursue clams, scallops, urchins, and most anything they can wrestle down. Many fish pursue scallops, crabs, lobsters, shrimp, and octopi. Some nudibranchs and a variety of snails graze on the algae that grow on the rocks. Other rocky area residents include barnacles, worms, and colorful fish such as blennies and rockfish. The cracks and fissures of the rocks provide ideal living quarters for many species.*

🐚 *Above: Often, a variety of colorful sea anemones cover the rocks in temperate waters. Many species of anemones reproduce by budding: Individuals essentially split themselves in half and become two separate animals that are genetically identical. Here, each of the two colors of anemones represents a genetically identical group. Members of one group are able to recognize members of the other as being different. Once they do, they will attack with special stinging cells in their tentacles and drive each other back. Eventually the battle will establish a characteristic anemone-free zone between the two groups.*

🐚 Opposite page: *This harbor seal is equally at home in the kelp forests or rocky areas of temperate water environments. Harbor seals cavort in the cool waters, feeding on fish that dwell in or near the forests. The seals, in turn, provide food for the great white sharks, killer whales, and other top-end predators that appear in the region. These transient predators frequently wander in and out of the shallow rocky areas in their endless search for food.*

🐚 Above: *A swell shark pup hatches from its egg case, which rests on the sea bottom. Several species of sharks that share similar behavior patterns and lifestyles commonly inhabit the rocky areas of temperate waters. Swell sharks, horn sharks, leopard sharks, and Port Jackson sharks typically grow to be only a few feet long. These small but effective predators stalk the rocky reefs in search of various crustaceans, mollusks, echinoderms, and small fish.*

🐚 Opposite page: *Hundreds of juvenile rockfish dart about a clump of drifting kelp in the deep waters of the Pacific. As juveniles, these rockfish stay in open waters. After they mature to a certain point, however, they will make their way to shallower areas over continental slopes and settle there. While in the open seas, they are preyed upon heavily and so form an important link at the base of the open-ocean food chain. Many other species of fish and invertebrates also move from one marine habitat to another at different stages in their development.*

🐚 Above: *A pelagic red crab in deep waters off Coronado Island, Mexico. Also called tuna crabs, these small crustaceans usually spend their time swimming freely in the open ocean. The crabs have been known, however, to invade coastal areas en masse—sometimes by the millions. Scientists do not know why the crabs make their way to shallow offshore waters. The behavior might be linked to reproduction, since many other species of marine crabs gather in large groups to spawn.*

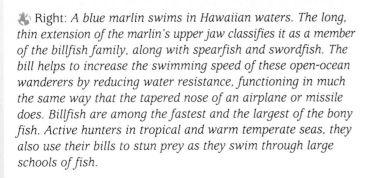 Above: *The ocean sunfish is a good example of the many unusual animals that make their homes in the open ocean. The larvae of the ocean sunfish are only a few millimeters long and look like tiny yellow pincushions. By the time they reach adulthood, however, they have grown into massive creatures measuring some six feet in length, weighing several thousand pounds, and bearing little if any resemblance to their modest beginnings. As adults, they feed on the huge numbers of jellyfish that drift aimlessly on the ocean currents.*

Right: *A blue marlin swims in Hawaiian waters. The long, thin extension of the marlin's upper jaw classifies it as a member of the billfish family, along with spearfish and swordfish. The bill helps to increase the swimming speed of these open-ocean wanderers by reducing water resistance, functioning in much the same way that the tapered nose of an airplane or missile does. Billfish are among the fastest and the largest of the bony fish. Active hunters in tropical and warm temperate seas, they also use their bills to stun prey as they swim through large schools of fish.*

🐚 Opposite page: *A pair of southern elephant seals (Mirounga leonina) bellow in the frigid waters of the Antarctic. The southern polar landmass is a region of ice, snow, and bitter cold. Temperatures commonly plunge more than 100 degrees F below freezing. The surface environment is so harsh that the biological cycle revolves around the marine kingdom. Life clings to the edge of the continent and a few other isolated inland areas where slightly milder conditions allow a handful of plants and animals to survive.*

🐚 Above: *This snowy white polar bear appears unfazed by the Arctic water's chilly temperatures. Polar bears serve as a link between land and sea in the Arctic realm. On land, they reign supreme, the undisputed rulers of the northern ice lands. Adult males can be 10 feet tall and weigh 1,200 pounds, making them the largest terrestrial predators in the world. Polar bears are also excellent swimmers, and they spend considerable time hunting in and around the Arctic waters. Like most bears, they are opportunists, taking whatever food they can find, but seals are a favorite part of their diet.*

🐚 Left: *A four-foot-tall emperor penguin shelters its chick. Emperors breed far inland at the height of the Antarctic winter, facing temperatures of 80 degrees F below zero and howling 100-mile-an-hour winds. Amid perhaps the world's worst weather conditions, the female lays a single egg and then departs for the sea to feed. The male carries the egg on top of his feet, covering it with a fold of abdominal skin. He will not leave the egg to feed for the entire two-month incubation period and can lose half his body weight before the egg hatches. About the time the chick emerges, the fattened female returns, and the male surrenders his charge and heads for the sea.*

🐚 Opposite page: *Emperor penguins crowd a snowy Antarctic bank, each waiting for its turn to enter the water and feed on fish and mollusks. Much jostling occurs at the water's edge, as the fate of the first few penguins to enter will indicate to the others whether a leopard seal or other predator is lurking in the icy waters. With their strong flippers, webbed feet, thick feathers, and layer of insulating fat, penguins are perfectly suited to their aquatic lifestyle. All 17 penguin species live south of the equator. Antarctic and sub-Antarctic species include gentoo, chinstrap, rockhopper, macaroni, and king penguins.*

🐚 Opposite page: *Even delicate anemones and sponges are able to thrive in Antarctica's frigid waters. In both polar regions, the changing seasons govern the cycle of life; sunlight is a crucial ingredient in this cycle. During the summer, sunlight is abundant and prevailing ocean currents and storms create deep-water upwellings that bring nutrients from the depths. These conditions nourish tiny marine plants called phytoplankton, which reproduce in immeasurable volume. Zooplankton, microscopic animals that feed on the phytoplankton, multiply rapidly as well. Together, these microscopic plants and animals underlie a food chain that supports sea stars, crabs, fish, birds, seals, and whales.*

🐚 Right: *Among the Arctic's many marine residents is the sea nettle. This jellyfish, which floats passively on the ocean currents, is one of the region's year-round inhabitants. Many of the larger Arctic and Antarctic animals are migratory. Drawn by the summer's rich food supply, they converge on the poles during the warmer months and then travel toward the equator during the harsh winter. Just as the summer's bounty brings life to the polar regions, the bitter winter drives much of it away.*

Right: *The deep-sea regions are home to some of the most bizarre, frightening-looking creatures imaginable. They are nature's answer to the unflagging darkness, intense pressure, and endless, cold expanse of the abyss. This fangtooth (Anoplogaster cornuta) does not frequently encounter prey in the isolated depths, but its formidable jaws ensure that few meals ever escape its grasp. The fangtooth is shielded from head to tail with hard, bony plates. The covering may be a barrier to predators, or it may serve to protect the animal from the crushing effects of the water pressure in the deepest reaches of the ocean.*

🐚 Above: *A deep-sea anglerfish attracts its prey with a biolu-
minescent lure that it wriggles from a long appendage on its
forehead. Many deep-sea species use bioluminescence in different ways for the basic necessities of life in the wild, such as
luring prey, attracting a mate, or confusing a predator. Another
common adaptation that this anglerfish enjoys is a widely
hinged jaw for swallowing large prey. Some deep-water species
also have distensible stomachs so that they can swallow whole
animals that are actually larger than themselves.*

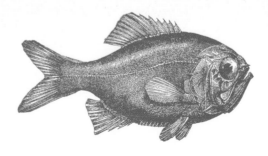

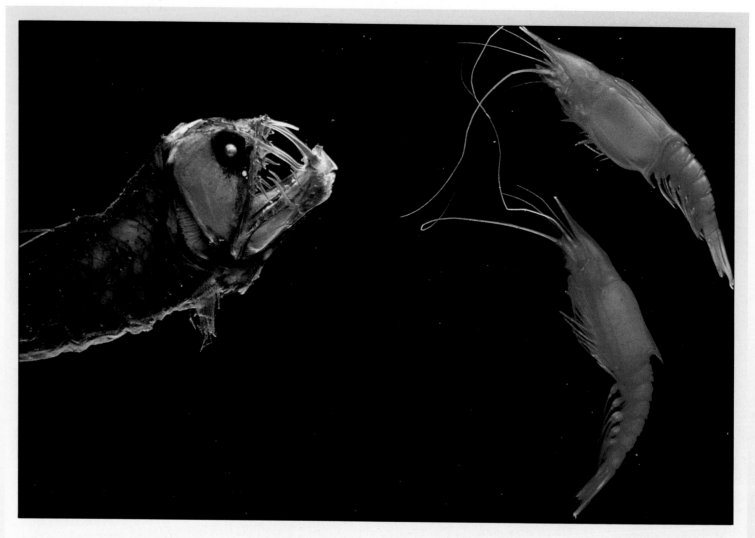

🐚 Above: *A viperfish encounters a pair of mysid shrimp in the depths of the ocean. In most open habitats, bright red makes a poor camouflage color for prey animals. In the depths of the ocean, however, it cloaks them from most other species. The thousands of feet of water above them filter out almost all of the red light from the sun, meaning that many predators will see the shrimp only as formless black shadows in the darkness.*

🐚 Opposite page: *A deep-water barnacle (Cirripedia scalpellidae) extends its feathery appendages to strain food from the nutrient-rich waters. Many deep-sea creatures subsist on the endless supply of decaying matter that rains down from above. When an organism dies in the surface waters, its remains often provide scavenging creatures at the bottom of the ocean with sustenance. While the deep-sea environment is harsh and life there sometimes depends on chance events like food falls, the ocean depths still support a rich and varied web of life.*

Invertebrates

AN INVERTEBRATE IS ANY ANIMAL lacking a backbone. Other than that, members of this group have very little in common. Some have sophisticated internal structures, and others are little more than a haphazard mass of cells. In the marine world, the term covers animals as different as the tiny coral and the enormous giant squid, or the primitive sponge and the highly developed octopus.

Invertebrates are the most varied and numerous life-forms in the sea. Representatives abound in every ocean environment, and they shape their habitats' life cycles in many ways. They serve as the primary food source for many higher animals, but they do much more. Corals, for example, literally shape the vista of their reef communities. Sponges feed on debris that would otherwise cloud the water and choke many species out of existence. And as a group, invertebrates add much to the beauty and wonder of the marine world.

SPONGES

Without muscles, nerves, and even specialized tissues, sponges are the most primitive multicellular animals. A sponge consists of several types of loosely gathered cells whose contribution to the whole organism is essentially coincidental. In their larval stage, sponges are free swimming, but most adults attach themselves to corals, rocks, or the shells of other animals such as hermit crabs and decorator crabs.

Worldwide there are about 10,000 species of marine sponges, and another 150 varieties live in freshwater. They occur in an incredibly wide variety of colors and shapes. Some sponges are symmetrical and others are irregular. Many simply take on the shape of whatever they happen to be growing on.

CNIDARIANS

Coral polyps and sea anemones are often called the "flowers of the sea" because their flowing tentacles resemble flower petals. But make no mistake, the similarity ends there. Corals and sea anemones bear deadly stinging cells concentrated in their tentacles. These creatures are close relatives of jellyfish and Portuguese man-of-wars, animals that even most non-specialists recognize as potentially harmful. All of these creatures, along with sea wasps, sea pens, sea pansies, and hydroids, are members of the phylum Cnidaria. Their most distinguishing characteristic is their stinging cells, called cnidoblasts.

Opposite page: *A brittle star rests on the top of a bright red sponge.* This page: *A pair of shrimp hide amid the polyps of a bubble coral colony.*

Inside each stinging cell is a structure called a nematocyst. Although nematocysts vary from one species to another, they all look somewhat like coiled harpoons. When stimulated chemically by the presence of potential prey or physically by touch or increased water pressure from even slight movements, the nematocysts fire and release deadly toxins.

MARINE WORMS

The two major divisions of marine worms are flatworms and segmented worms. Flatworms are exactly that—flat. Most of the 4,000 or so marine species are less than two inches long. However, several tropical species exceed six inches in length. Many marine flatworms are quite colorful, and they swim in a series of rhythmic undulations that divers often find captivating.

Segmented worms are far more advanced than flatworms. They have a more advanced brain, and they have specialized sensory organs concentrated mainly near the head. Segmented worms are also known for their delicate beauty. The exposed part of the worm is a colorful, intricate group of branched tentacles called radioles that strain plankton from the water. The rest of the animal, which looks much like an ordinary worm, lives burrowed in the coral, rock, or sand, or in a self-made tube.

BRYOZOANS

The word bryozoan is derived from two Greek words; *bryo* means moss, and *zoan* means animal. In some respects bryozoans look quite a bit like small patches of moss. There are more than 4,000 species of bryozoans around the world, but scientists know little about the animals as a group.

The individual animals, called zooids, are so small that it takes a microscope to distinguish them. Each animal consists of an elongated body topped by a ring of tentacles that surround the mouth. They lack specialized

organs for respiration, circulation, and excretion, but within a colony, various individuals are specialized in their roles. For example, some individuals are equipped with tiny jaws to discourage intruders.

MOLLUSKS

The phylum of mollusks contains such seemingly diverse animals as snails, octopi, scallops, and nudibranchs. As unlikely as it seems, evolutionary specialists believe that all mollusks arose from a single group of common ancestors. Today approximately 75,000 species of mollusks inhabit the Earth, and representatives are found in every marine habitat. Mollusks are well-developed animals, having distinct organ systems and sophisticated senses. In fact, many experts believe that octopi are the most intelligent of the marine invertebrates.

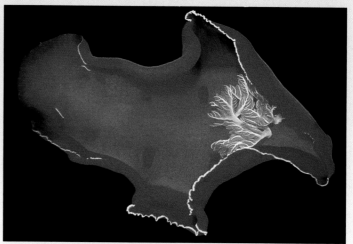

Above: *This nocturnal Spanish dancer nudibranch is one of the many mollusks found in the waters of the Red Sea.*

Mollusks share a number of common characteristics, many of which are unique to members of this phylum. All have a soft, fleshy body called the visceral mass; all have a mantle that usually secretes a shell; and all have some type of muscular foot that is used in locomotion or in digging. Snails use their foot to walk or crawl, clams dig with their modified foot, and even octopi crawl or walk across the seafloor on a modified foot that has evolved into eight tentacles. Many, but not all, mollusks have an external shell.

CRUSTACEANS

More than one million species, over 75 percent of all animals, belong to the phylum Arthropoda. Most are land-based insects, but many arthropods play major roles in the ocean environment. Marine arthropods such as lobsters, crabs, shrimp, and barnacles are known as crustaceans.

Above: *Decorator crabs are among the more brightly colored species of crustaceans.*

All crustaceans live inside a hard shell called an exoskeleton, a feature that provides both advantages and disadvantages. The hard shell offers valuable protection, and it gives muscles a place to attach so that movement is possible. The primary disadvantage is that for growth to occur, the animal must go through a molting process that sheds the old shell and forms a new, larger one. Molting can take from several hours to nearly a full day, and the animals are left very vulnerable to predators during the process. Without their shells, most crustaceans are virtually defenseless.

ECHINODERMS

All echinoderms live exclusively in the marine environment. The group includes sea stars, brittle stars, basket stars, sea urchins, sand dollars, sea cucumbers, and crinoids. The name echinoderm means spiny-skinned, and it fits them well. Most have obvious spines or wartlike projections, and all display five-sided, or pentamerous, symmetry. The symmetry is very easy to see in sea stars and brittle stars. Most species possess five arms that radiate from a central disc. At first glance, sea cucumbers and sea urchins seem an exception to the rule, but their bodies do have five identical segments.

Echinoderms have a skeleton of spines or hard plates. They also have distinct organ systems, making them the most advanced marine invertebrates. A unique water vascular system controls the hundreds of tube feet that allow them to move. This hydraulic system also assists some species in capturing prey.

Modern echinoderms are bottom dwellers. They roam the seafloor in both shallow and deep waters searching for food. Most live in temperate seas, but some make tropical waters their home. Echinoderms range in size from sea cucumbers only a half inch long to sea stars over three feet wide.

🐚 *Sponges come in a variety of shapes and sizes and in all colors of the rainbow. Left: Bright yellow tube sponges grow in clusters and reach a height of more than three feet. Above: Cup sponges come in brilliant shades of orange and red. Opposite page, top: The red rope sponge has a distinctive growth pattern that often leaves it looking like a tangled coil of heavy rope. Opposite page, bottom: These beautiful purple tube sponges are common in the Caribbean and Dutch Antilles. Opposite page, right: This azure vase sponge clearly shows the pores and channels that are typical of the sponge family.*

Left: *Porifera, the scientific name given to the phylum of sponges, refers to the canallike system of pores, channels, and openings in their bodies that allows a steady current of water to flow through the sponges. Sponges draw oxygen and filter food particles from the water as it passes through them. The water exits through large openings called oscula and takes waste products with it as it goes. Sponges play a major role in the ecology of their habitats by filtering bacteria, debris, and other particles from the water. This large basket sponge is spawning by releasing a cloud of sex cells with its exiting water current.*

Opposite page: *A fish seeks refuge behind a sponge. Sponges provide security for many animals. Brittle stars, worms, shrimp, crabs, juvenile lobsters, and snails are among those that use the hidden recesses of sponges as places to hide. Sponges have few natural enemies. Their foul taste and noxious secretions make them painfully unappetizing to most animals. Still, some species of sea turtles, angelfish, filefish, and sea slugs do prey upon them.*

🐚 *Tropical hard corals come in a dazzling array of forms. The common names of the animals shown here evoke the seemingly endless variety of shapes and growth patterns found in the reef-building corals. Above:* antler coral; *above left:* leather coral; *left:* brain coral. *Opposite page, left:* elkhorn coral; *top right:* fungus coral; *bottom right:* table coral. *Each of these structures is actually a colony that contains hundreds of individual animals called coral polyps. The polyps are all encased in hard calcium carbonate skeletons that fuse together to form the larger structure.*

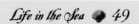

This page: *Sedentary hard corals depend on their tentacles, armed with stinging cells, to capture plankton that drifts by in the currents. When feeding (below), the polyps appear flowerlike as they open up and extend their tentacles into the water column, waiting to sting any unsuspecting prey. When not feeding or when threatened (right), the tentacles retract, giving the animals an entirely different outward appearance. The typical coral polyp is tube shaped, with a ring of tentacles surrounding a mouth at the top.*

Opposite page: *The skeletal remains of hard corals that form large branching or mounding colonies serve as the landscape in the habitats we call coral reefs. Each species of coral contributes its own unique growth pattern to create the structure of a reef. Fish, crabs, octopi, sharks, sea stars, sponges, and countless other organisms find a home there. They hunt, flee, forage, and hide in and around the complex maze of crevices, passageways, and caves that have been formed over millennia by the coral colonies.*

Opposite page: *The individual polyps of this gorgonian coral sea fan give it a soft, fuzzy outward appearance. In contrast to hard corals, gorgonian or soft corals are not reef builders; their bodies do not develop the calcium carbonate outer skeletons that the hard corals do. They are like hard corals, however, inasmuch as they occur in colonies consisting of hundreds of individuals. Soft corals include the animals commonly referred to as sea fans, sea whips, and wire corals.*

Right: *A sea fan (gorgonia flabellum) waves back and forth in the warm Florida waters. As it does so, it traps passing prey and soaks up the warm sunlight with its branched tentacles. Sea fans live in tropical and temperate seas around the world. The colors of soft corals range from the bright pinks seen here to cherry red, burgundy, yellow, orange, and even pure white. Large colonies can stand taller than a person.*

🐚 Opposite page: *Sea anemones have much in common with their close cousins, the corals. The basic body structure of the individual animals is quite similar. One noticeable difference, however, is that the anemone is a great deal larger. In temperate waters, individual sea anemones can be baseball-size; in tropical waters, they can be more than three feet in diameter. Also anemones are not colonial animals like the corals. Many live a solitary existence and some may congregate in clusters or groups, but they do not form true colonies.*

🐚 Above: *A rose anemone feasts on a blood star. Many anemones feed on plankton, but larger species will readily consume small fish, sea stars, or most any organism they can overcome. Like corals, sea anemones are simple feeding machines. The tentacles wave back and forth in the water column waiting to encounter a meal. When an organism wanders too close, the tentacles firmly take hold and fire their stinging cells. The subdued prey is then drawn down toward the single opening at the center of the tentacles and consumed.*

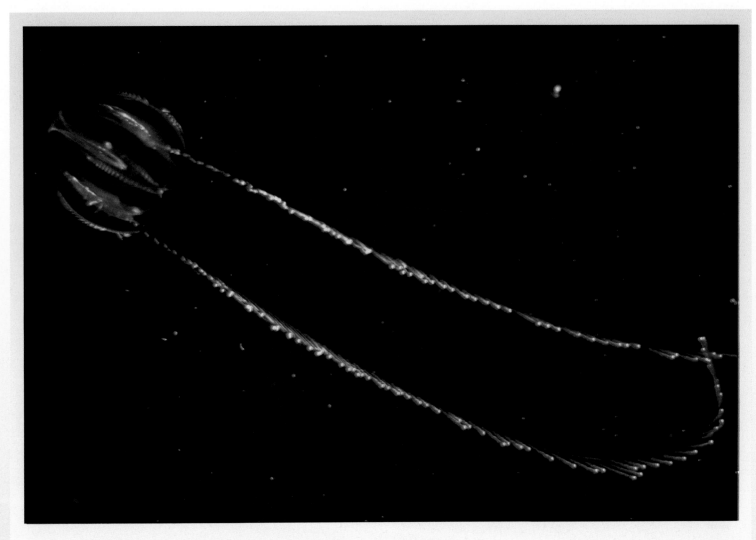

Above: *Jellyfish are fairly poor swimmers and usually go wherever wind and currents push them. Because they can't actively hunt their prey, they depend on unwary or curious animals coming into contact with their tentacles. The tentacles of many species are nearly transparent and can trail 100 feet behind them. The bright appendages on the tentacles of this comb jelly act as lures to potential prey; small fish will mistake them for the free-floating organisms they often feed on, and when the fish move in to investigate, they become ensnared in the comb jelly's tentacles.*

Opposite page: *A Portuguese man-of-war ensnares its next meal. The Portuguese man-of-war is not actually a true jellyfish. Similarities in body structure relate it more closely to the hydroids, another group of cnidarians, than to jellyfish. Man-of-wars are actually colonies of individual animals in both the medusa (free-swimming) and polyp (stationary) stages. Their iridescent blue to purple bells, or floats, catch the winds and currents at the surface. Below and behind the bell trail a number of very potent tentacles that can easily kill prey such as mackerel and anchovies.*

Above: *The sea pen, like its cousin the sea pansy, is a form of colonial cnidarian that lives in the sand. Each pen is decorated with numerous little polyps. When feeding, the polyps extend and give the colony a soft, bushy appearance; when threatened, the polyps retract, and the sea pen looks more like a dead leaf. In some species, the stalk itself will retract into the sand to avoid a threat. Right: This feathery, branched hydroid colony looks much like a small bushy plant. Although most hydroids are colonial, some are solitary. Hydroids often attach to the end of sea fans and sea whips, where their tentacles can reach into the water column and snare drifting plankton.*

🐚 Opposite page: *Many free-living marine flatworms are remarkably colorful, and their bodies are very thin. Closely related to terrestrial tapeworms, most marine flatworms are bottom dwellers, and some species are parasitic. Flatworms are the most primitive animals to display bilateral symmetry; their bodies have distinct left and right sides that mirror each other. Many flatworms are cannibalistic, yet many reproduce sexually, a combination that presents a rather interesting problem. During the breeding season, individuals need to be able to identify potential mates that will not eat them.*

🐚 Above: *A fireworm, or bristle worm, makes its way across a gorgonian coral in the waters off Palm Beach, Florida. Though small, the bristle worm has formidable, spiny defenses: each bristle, or setae, is hollow and filled with venom. If touched, even ever so slightly, its fine bristles will pierce the offender and often cause severe pain and swelling. Segmented worms like the bristle worm have a body composed of a series of identical segments joined together; the head region is separate and contains the mouth and several sense organs.*

Right: *The whorled gills of a Christmas tree worm extend out into the water for respiration and for filtering out small food particles. Among the most beautiful and delicate of all marine creatures, many segmented worms bear little resemblance to the terrestrial worms people know and often loathe. Spiral gill worms like this one build tubes down into hard coral colonies. A single coral may be peppered with many worms. If removed from their tubes, the bodies of these marine worms look very similar to those of their terrestrial relatives.*

Opposite page: *The feather duster worm will retract its frilled gills into a self-made tube at the slightest provocation. However, a patient observer will be rewarded as these cautious animals poke back out. Another interesting segmented worm is the bloodworm. Equipped with powerful jaws and fangs, bloodworms are voracious predators. When a bloodworm detects motion near the opening of any of a series of tunnels it builds in the muddy bottom, the worm quickly seizes and devours its prey, mainly crustaceans and other invertebrates.*

🐚 Opposite page: *This rose coral, or rose-de-mer as it is sometimes called, is not a coral at all but a bryozoan. Bryozoans attach themselves to sand, rocks, kelp, coral skeletons, and even other animals. Here, a bryozoan colony provides sheltering crevices for several small sea crabs. Bryozoan colonies range in color from drab brown, to snow white, to soft pinks and light tans. Although many colonies are fairly small— only a few inches across—some can be several feet wide.*

🐚 Above: *Bryozoans, also called moss animals, will grow in patches on almost any underwater surface. Bryozoan colonies take on a variety of shapes. Colonies can be branched and erect, or encrusting. Individuals tailor their body structure to benefit the group. Inconspicuous by size, bryozoans are still remarkably pretty animals. Many bryozoans are hermaphrodites, with each individual being both male and female at the same time. In some cases, their sex cells are produced alternately, and at other times, the same animal simultaneously produces sperm and eggs.*

🐚 Above: *Many snails graze on algae, using a rasping tonguelike organ called a radula; some snails, however, are active predators, devouring mainly other mollusks. This highly venomous cone snail is one of the most active carnivorous snails. Found in the tropical and subtropical waters of the Indo-Pacific and the Atlantic, these snails inject potent toxins into their prey with a detachable radular tooth. Because their shells are highly attractive, some people like to collect cone snails, but several Indo-Pacific species carry a venom that can be fatal to humans.*

🐚 Opposite page: *The largest and most diverse class of mollusks, the gastropods, or stomach-foots, includes the snail, abalone, conch, and cowrie. Gastropod shells occur in a wide variety of patterns and shapes, although the color and structure are consistent for any given species. Some of the most beautifully patterned shells are those of the cowries, such as this flamingo tongue cowrie. The cowrie frequently spreads its body tissue, or mantle, across the shell, leaving it with a glossy, highly polished finish. The mantle, which often has a colorful pattern, consists of two lobes that the cowrie can draw up on either side of the shell until they meet in the middle; the mantle can also withdraw all the way into the shell.*

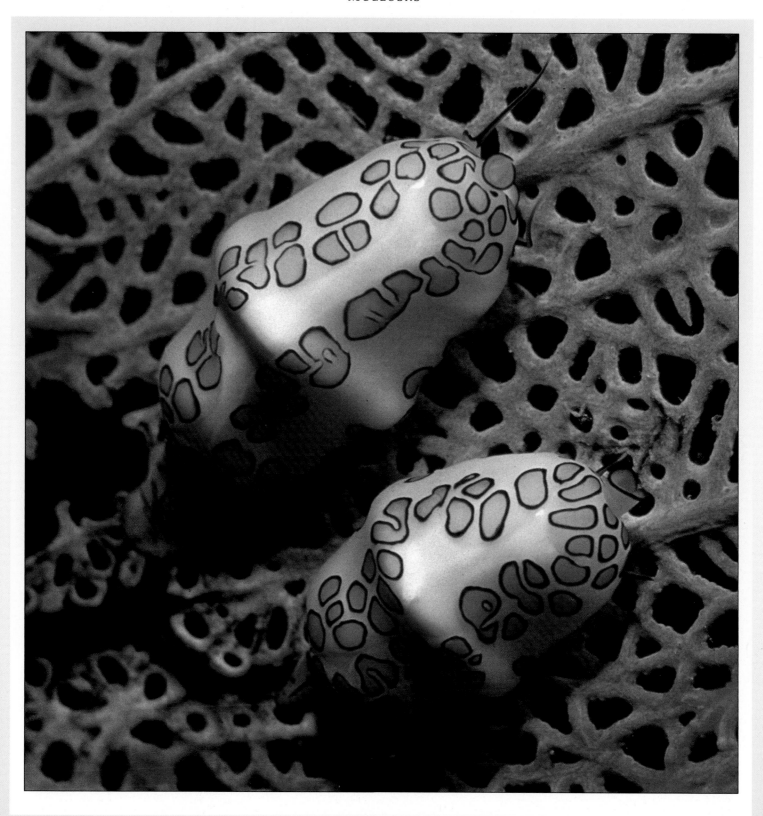

Above: *By exuding its mantle across the top of its shell, this limpet is able to escape from the clutches of a predatory sea star. The underside of the sea star's arm has numerous tube feet that can effectively grasp and hold prey. The limpet's mantle may be too slippery for the tube feet to take hold. The limpet may also secrete some substance that repulses the attacker, because sea stars will often break off an attack immediately upon contacting the mantle.*

The soft, compressed bodies of clams, scallops, and oysters reside inside two hinged shells that can be clamped tightly shut by a tough ligament. Above: Many clams are small creatures that spend their adult lives buried in the sand or mud, extending tubelike siphons into the water to draw in food and oxygen and expel wastes. One notable exception is this giant clam of the Indo-Pacific. These clams can reach a length of three feet or more and can weigh 400 pounds. Giant clams begin their lives in free-swimming larval stages and then settle on top of the seafloor as adults. Right: Light-sensitive eyespots and short tentacles border the edges of this temperate-water scallop's shells.

🐌 Above: *This frilly specimen, commonly called a lettuce-leaf nudibranch, bears testament to the bizarre shapes and lovely colors for which nudibranchs are known. The name nudibranch derives from Latin terms meaning naked gills: Nudibranchs lack the external shells found in most mollusks, and so their gills are exposed. Ranging in length from less than one-half inch to nearly two feet, nudibranchs are important members of reef communities throughout tropical and temperate waters.*

🐌 Opposite page: *Perhaps one of the most familiar species of nudibranchs is the Spanish dancer. Approaching 24 inches, Spanish dancers are among the largest species. The name comes from their habit of swimming over and around their reef habitats in a captivating motion reminiscent of Spanish flamenco dancers. Some nudibranch species spend most of their time endlessly grazing on algae-covered rocks, while others are carnivorous; the carnivorous ones prey on a variety of small invertebrates, including other species of nudibranchs.*

Above: *These two sea lemon nudibranchs have met, mated, and laid their spiral strands of eggs next to each other. Each strand of the mass contains thousands of eggs, many of which will end up as food for other animals. Right: When threatened, sea hares are able to release a noxious ink produced by specialized glands. Like their close relatives the nudibranchs, individual sea hares each possess both male and female reproductive organs. Sea hares differ from nudibranchs in that they have a pair of swimming fins that fold over their backs when not in use. Sea hares live in tropical and temperate waters and graze on algae as their primary food source.*

Left: *A rainbow nudibranch feeds on a tube anemone. Although the anemone's tentacles contain highly sensitive stinging cells that drive away most other predators, the nudibranch is able to attack with impunity. It does this by erecting clusters of elongated structures called cerata that make its victim perceive it as another anemone. The nudibranch then feeds on the anemone's tentacles, biting off pieces with its bladelike jaws. The tentacle pieces are swallowed intact, and the stinging cells somehow remain undischarged. Special ciliary tracts within the nudibranch's body move the ingested stinging cells to the tips of the cerata on the nudibranch's back. Once there, they become a defense for the nudibranch; should a predatory fish try to make a meal of the nudibranch, the stinging cells will discharge into the attacker.*

🐚 Above: *A cuttlefish swims slowly along, propelled by a fin that runs around its body like a short tutu. When necessary, these jet-propelled animals can move swiftly. Like their close relatives squid and octopi, cuttlefish are cephalopods, or head-foots. All species have a prominent head and attached arms or tentacles equipped with suckers. Cephalopods are among the most advanced and intelligent marine invertebrates. Their sight and sense of touch are especially well developed.*

🐚 Opposite page: *The squid's large, glassy eyes, splayed tentacles, and translucent body make it easy to identify, even from a distance. One species of squid, the giant squid, is the largest of all invertebrates. It has been documented at lengths of up to 45 feet, although specimens are rare because these animals live in extremely deep waters. In squid, as in all cephalopods, the sexes are separate. Squid often spawn in large groups; in many species, the adults die soon after they mate.*

Above: *The giant Pacific octopus inhabits rocky shores off California, Oregon, Washington, and Canada. These animals are quite shy by nature, but they are large enough to command any diver's undivided attention. Forty to fifty pound specimens are common, and some weigh in at 600 pounds and measure 16 feet in length.* Right: *A young octopus emerges from its egg. Octopi lay eggs in clusters that look like bunches of grapes. For a month or longer, the females guard and diligently clean the eggs until they hatch; they do not feed while tending the eggs, and they usually die shortly after the young are born.*

 Above: *As a rule, cephalopods are voracious predators. They have sharp, parrotlike beaks, and many species inject venom when they bite. The bite of this blue-ringed octopus can be fatal to humans. Like other cephalopods, octopi also have glands that produce an inky substance they can release for defensive reasons. The ink clouds the water, obscuring the octopus and confusing any would-be predators; the ink may also block the chemical receptors that some predators, such as moray eels, use to hunt.*

Right and below: *Although they spend most of their time dwelling on the bottom, using the rows of suckers on their eight arms to move easily along the seafloor, octopi are excellent swimmers when they have to be. Like squid and cuttlefish, they can swim very quickly by taking water into an area called the mantle cavity and then rapidly forcing the water out a directable tube called a siphon. The force created by expelling the water propels them in the opposite direction.*

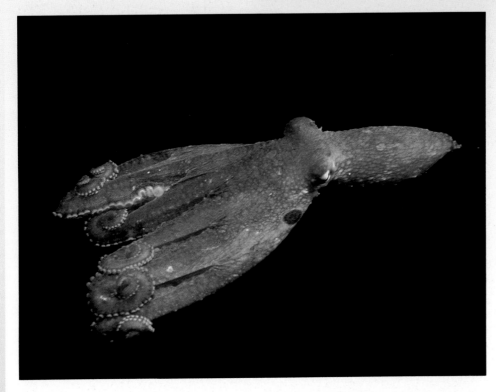

Opposite page: *Octopi have a remarkable camouflage ability. They can quickly change the color and even the texture of their skin to create an unimaginable variety of exterior patterns to match almost any background. An octopus can be dark brown and smooth one minute, cherry red and knobby the next, and only a moment later be bright blue and ruffled. Lacking the hard shell found in most other mollusks, soft-bodied octopi are also very adept at escaping predators; even large specimens can slip to safety through very small openings.*

Right: *Not all lobsters have large claws like the familiar New England lobster. Some species, like this spiny lobster, have a number of sharp spines instead. Like all the decapod crustaceans, lobsters have ten walking legs. They are generally shy and retiring during the day; under cover of darkness, however, they roam the bottom looking for food—mostly fish, mollusks, and other invertebrates. Lobsters often gather together in caves or crevices for shelter; it is not uncommon to find dozens of lobsters wedged into a tiny crevice.*

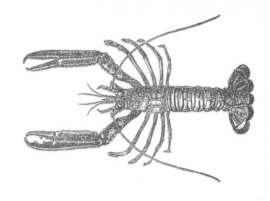 Above: *The usual long antennae are modified into broad, flattened plates on the head of this slipper lobster. Slipper lobsters are a part of many coral reef, rocky reef, mud, sand, and grass communities. In most situations, lobsters move by walking along the bottom on their ten jointed legs. When they are threatened by one of their many predators, though, they can move quickly for short distances by rapidly pulling their muscular tails downward beneath their bodies to propel themselves backward through the water.*

🦐 Above: *An arrow crab sits motionless on a lacy purple sea fan. With its long, jointed legs, the spidery arrow crab may look awkward and ungainly, but it can be surprisingly agile and quick as it scurries around its habitat. As a group, arthropods show amazing diversity and specialization in their appendages. A particular species' body structure can reveal significant behavioral information, such as where it makes its home or where and what it prefers to hunt.*

🦐 Opposite page: *Some crabs, such as spider crabs, blue crabs, and king crabs, live completely covered by their hard exoskeletons. However, only the front portion of this hermit crab enjoys that kind of protection; its soft abdomen is exposed to attackers. To protect themselves, hermit crabs crawl inside the discarded shells of snails. The newfound armor protects the hermit crab's vulnerable body parts. When a hermit crab outgrows its acquired shell, it will actively search out and move into a larger one. A small hermit crab will sometimes take the discarded shell of a larger hermit crab.*

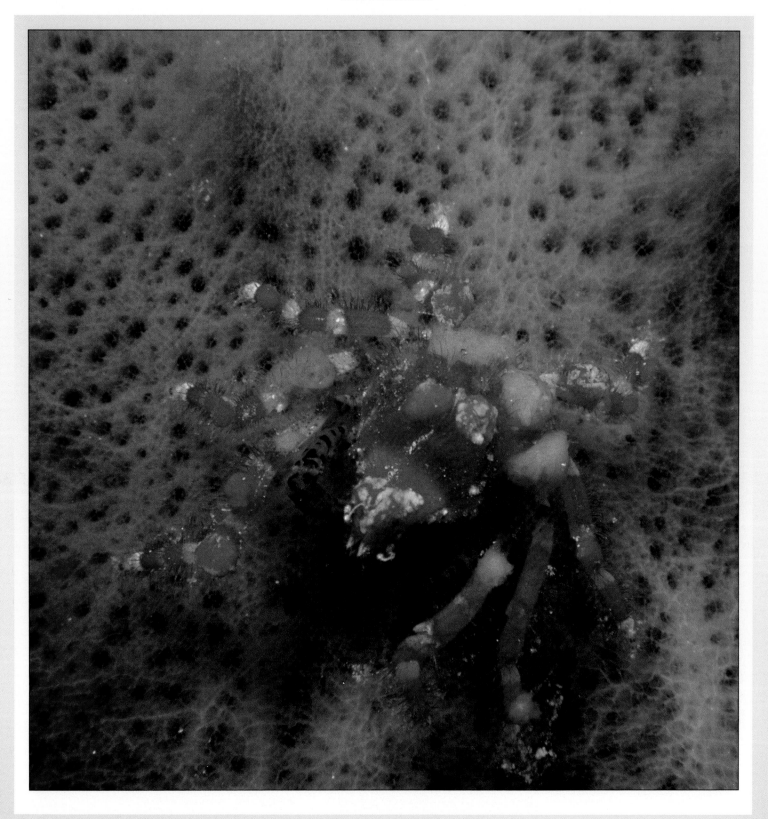

🐚 Opposite page: *Several species of crabs cover their outer skeletons with sponges, anemones, hydroids, algae, and other living organisms as a means of camouflage. Some of the decorator crabs bear hooked hairs, or setae, on their exoskeletons, and the hairs serve as an anchor for the organisms that attach to the shells. These organisms actually live and grow on the shell of the crab. If the crab remains still, predators are likely to interpret it as a part of the substrate rather than as a potential meal.*

🐚 Above: *This hermit crab off the coast of Hawaii carries several anemones on its shell. In this case, the decorator crab's adornments act more as a deterrent to predators than as camouflage; the anemones' tentacles will inflict painful stings on almost anything that comes in contact with them. The normally sedentary anemone benefits from the arrangement by taking advantage of the crab's mobility; its tentacles capture any small planktonic organisms that stream through the water as the crab scurries about.*

🐚 Opposite page: *Worldwide, there are more than 2,000 species of true shrimp. All have ten walking legs; their elongated bodies consist of a segmented tail and a head and thorax that are fused together. Some shrimp are excellent swimmers. For bursts of speed over short distances, they rapidly pull their tail underneath their body and propel themselves backward in much the same way that lobsters do. At night, shrimp are one of the more common groups of animals in most marine settings.*

🐚 Above: *In many reef communities, pistol shrimp like this one are quite common. At night, the crackling and popping of these shrimp as they snap their claws shut can sometimes be heard all across a reef. The snapping is accomplished by rapidly contracting a disclike muscle near the claws. The powerful sound momentarily stuns potential prey and allows the shrimp to capture it. Another predatory shrimp, the mantis shrimp, uses its razor-sharp claws and quick reflexes to impale small fish and crack open hard-shelled prey, such as mollusks.*

Opposite page: *Very few animals will ever willingly venture this close to the mouth of a predatory eel, but this cleaner shrimp boldly removes parasites, dead tissue, and bacteria from the eel's skin knowing that the eel will not harm it. Many species of shrimp provide such cleaning services to a variety of fish. The shrimp may use their claws to dislodge stubborn parasites; sometimes they even make incisions to remove invaders from beneath their clients' skin. Pederson's cleaning shrimp are prominent cleaners in Caribbean waters, while red rock shrimp do the job in reef communities along North America.*

Left: *The long, white antennae protruding from the top of this sponge are a clear sign that some coral-banded shrimp have taken up residence. These cleaner shrimp typically rock back and forth, waving their antennae as a sign that they are looking for someone to clean. Cleaning stations such as these often become areas of very high traffic. Often the fish will literally line up around the cleaning station and wait for a turn to be serviced.*

🐚 Opposite page: *Barnacles are an atypical group of crus-taceans, both in their appearance and in their behavior. Like many crustaceans, they are free swimming in their larval stage. As adults, however, they secrete a strong, bonding adhesive to attach themselves permanently to solid surfaces such as rocky reefs, corals, boats, pier pilings, logs, and even whales, turtles, and other living animals. Once attached, a barnacle extends its feathery cirri, which filter the water for food particles and then draw them down to its mouth.*

🐚 Above: *A parasitic isopod has taken a firm hold on this French grunt. Isopods, like their smaller cousins the copepods, are often parasitic, living on or in various fish and inverte-brates. Isopods are visible with the naked eye, but many cope-pods are microscopic, floating as plankton and feeding on small plants, animals, and decaying organic matter. Krill, a related group of crustaceans, are usually only one or two inch-es long. A surprising number of creatures depend on krill as a major portion of their diet—numerous invertebrates, some birds and seals, and even many of the great whales.*

🐚 Above: *A delicate lavender crinoid sits atop a sea fan. The oceans once housed several thousand species of crinoids; today their numbers have dwindled to about 600. Crinoids are the most primitive form of echinoderms. They lack both the spines and the small, pincerlike organs called pedicellariae found in most echinoderms. As in all echinoderms, though, their bodies have five symmetrical segments, and they have a great number of tube feet that they use for gathering food.*

🐚 Opposite page: *Crinoids can sport brilliant hues of burgundy, red, yellow, green, and an almost endless variety of mottled color combinations. Their long appendages quickly fork several times and form branches; these branches support numerous thin tube feet that collect plankton and other food from the passing currents. When not feeding, they keep their arms coiled inward. Some crinoids, known as feather stars, have small clawlike tentacles they use both to move and to hold themselves in place; others, known as sea lilies, are attached to the seafloor by stalks.*

Opposite page: *A group of spiny purple sea urchins sprawl across a rocky seascape. All sea urchins have spines of one sort or another. Some species have long, thin, razor-sharp spines; others have dull, stubby spines. Some species of irregular urchins, such as heart urchins, sea biscuits, and sand dollars, have shorter spines and a slightly different body structure. Most sea urchins inhabit solid or rocky environments, but the irregular urchins often prefer sandy sea beds where they can bury themselves.*

Above: *Sea urchin spines are quite robust in some species, meant more to deter than to injure predators. Though the urchins' spines provide a formidable defense, some fish, crabs, snails, other echinoderms, and sea otters still prey on them. Most urchins themselves feed on algae or on kelp plants. Some are scavengers, eating almost anything they come across. Some will even attack other species of urchins. An urchin's mouth is located on the soft underside of the body.*

Left: *When disturbed, sea cucumbers can eject a mass of spaghettilike tubules, which sometimes contains toxins, to ward off their agitators. Many sea cucumbers actually expel some of their own internal organs as a way of escaping predators. The strategy works because of the sea cucumber's amazing regenerative ability, which allows it to quickly replace the lost organs. With their sausage-shaped bodies, sea cucumbers don't seem to have the same structure as other echinoderms. Their elongated bodies do, however, have the five symmetrical sections that make the echinoderms distinctive.*

Opposite page: *Sea cucumber coloration ranges from drab brown or dark green to snowy white or bright orange. Knobby wartlike projections or spines frequently cover the body. Most sea cucumbers are either deposit feeders or suspension feeders. The deposit feeders use their adhesive tentacles to pick up decaying organic matter wherever they can find it. The suspension feeders, like this one, extend their branched tentacles into the water and collect food particles.*

🐚 Opposite page: *More than any other echinoderm, sea stars clearly show the five-sided symmetry that characterizes the group. Slow-moving sea stars have hundreds of tube feet on the bottom of their flexible arms that they use to make their way across the irregular seafloor. Most sea stars have five arms that grow out from a central disk, but some have as many as 40 arms. Those with more than five arms are commonly named sun stars. All sea stars display remarkable regenerative abilities, allowing them to regrow lost arms.*

🐚 Above: *A cushion star, or shark's pillow as it is sometimes called, is an unusually full-bodied sea star. Many sea stars forage for debris as they slowly creep across the seafloor. A number of other species are relentless hunters, feeding on snails, scallops, clams, crustaceans, segmented worms, and even small fish. One species, the crown-of-thorns sea star, consumes reef-building corals in shallow tropical waters. Very often these pervasive predators can have a strong impact on the entire life cycle of their habitats.*

🐚 Above: *A sea star uses its powerful tube feet to pry open the shell of a clam. In some instances, a sea star will actually extrude its stomach outside its body to get a meal. When a predatory sea star finds a clam or scallop, it wraps its arms around the bivalve's shells and grabs each half with its tube feet. If the sea star is unable to force the shells apart, it can extrude its stomach through a tiny space between the shells and secrete enzymes that will digest the prey inside its shell.*

🐚 Opposite page: *A brittle star sits atop a vase sponge in the waters off Honduras. Brittle stars have long, usually spiny arms, and their central disks are much more obvious than those of sea stars. Each arm of a brittle star consists of many separate segments that allow the arm to move in a sinuous, snakelike manner, making the creature fast and very maneuverable. Some brittle stars inhabit coastal areas and tidal pools, but most species prefer deeper water. Some brittle stars actively hunt, while others scavenge for a living.*

Vertebrates

APPROXIMATELY 46,000 SPECIES of vertebrates inhabit the Earth today. They all possess a skeletal system including a backbone or spinal column, well-developed organ systems, and well-developed senses. This familiar group includes fish, reptiles, mammals, birds, and amphibians.

In the marine world, vertebrates are among the most obvious creatures. They tend to be larger and more active than most other ocean dwellers, and often they dominate their habitats. Marine vertebrates include snakes and turtles, sharks and countless fish species, and whales, dolphins and seals. They thrive in literally every ocean environment.

TUNICATES

Tunicates, or sea squirts, are an unusual group. Worldwide there are more than 1,300 species, and they occur in all seas. Tunicates are an evolutionary link between invertebrates and vertebrates.

Tunicate larvae resemble small tadpoles, with an enlarged head tapering to a thin tail. A notochord—a flexible mass of cells running the length of the body and giving structural support—is present in the tail. Adults, which have lost both tail and notochord, are tube or barrel shaped. They are encased in a jellylike or leathery case or tunic, which gives them their name.

MARINE REPTILES

Of the 7,000 reptile species in the world today, less than 70 are marine species. Aquatic reptiles probably evolved from their terrestrial cousins. They have lost many adaptations found in their land-dwelling relatives but have developed other features for survival in the sea. All reptiles are cold-blooded, and most marine reptiles require the warmth of tropical or temperate seas in order to survive.

The marine iguana is the world's only sea-going lizard. Though fierce looking, it is actually quite shy. Found in the Galápagos Islands off Ecuador, it spends most of its time on rocky shores but enters the sea to graze on algae each day.

All species of sea snakes are native to shallow Indo-Pacific waters, where they feed on fish. Species such as the yellow-bellied sea snake display attractive color patterns, although others, such as the olive sea snake, are drab green or light brown. All are excellent swimmers. Their tails have developed into broad, flat paddles that propel them through the water. Sea snake venom is considerably more potent than that of terrestrial snakes, but sea snakes show little unpro-

◄⧮► This page: *This gorgeous solitary tunicate is attached to the seafloor in the Indo-Pacific.* Opposite page: *A green sea turtle perches on a reef off Hawaii.*

voked aggression toward humans, except during mating season. Like all reptiles, sea snakes are air breathers.

A third type of marine reptile is the sea turtle, found in tropical and temperate seas around the world. Sea turtles differ somewhat from their land-based cousins. Their limbs and head cannot retract into their shells, and their flippers have evolved from legs to assist in swimming. Sea turtles are excellent swimmers, and though they usually cruise quite slowly, they may be capable of short bursts exceeding 20 miles per hour. Sea turtles are also superb divers. Leatherback turtles, for instance, can dive as deep as 3,300 feet. Typical of cold-blooded creatures, sea turtles have slow metabolic rates, a feature that allows them to dive under water for as much as 45 minutes at a time.

BONY FISH

Scientists have currently documented more than 20,000 species of fish, and some maintain that at least 8,000 species remain unclassified. All fish have gills throughout their lives, and most have fins. Many have scales, and all secrete mucus from their skin that blocks bacteria and fungus and reduces water friction.

Fish have well-developed eyes that are very similar to the eyes of humans, and they have well-developed inner ears (though they have no outer or middle ear). Surprisingly enough, fish depend heavily on their sense of smell. Olfactory receptors line numerous openings along the snout. Some fish emit chemicals with distinct odors to communicate with members of their species.

Fish vary in size from the 12-millimeter Marshall Island goby to the 40-foot whale shark, the largest living cold-blooded animal on Earth. Species such as puffer fish have inflatable bodies, while the inflexible bodies of sea dragons and sea horses rest in an armored coat of rings and plates. Pipefish, trumpetfish, and needlefish have long thin bodies. Ocean sunfish are large oval discs.

Above: *When threatened, this puffer fish takes in large quantities of water to expand its body and erect its spines.*

Feeding strategies vary considerably among fish species. Some fish graze on the large patches of algae that grow on coral reefs or rocky surfaces. Bottom-feeding trunkfish often blow jets of water into the sand to uncover buried prey. Filter-feeding anchovies and herring have mouths that strain food from the water. Trumpetfish and pipefish have elongated mouths that suck in prey. Predators like the barracuda and tuna snag prey with their sharp teeth.

 Above: *A group of surgeonfish graze on a large patch of algae growing on the substrate.*

Fish species also vary in territorial instincts. Many simply wander through the sea throughout their lives; although they generally prefer certain areas at given times of year, they do not defend a territory in which they feed or breed. Some of these species, like jacks, tuna, mackerel, anchovies, and herring, live in large schools, while marlin and sailfish are nonschooling species that roam the open ocean. Like their oceanic relatives, the schooling fish in reef communities do not claim territories. However, solitary reef-dwelling species, such as damselfish, do claim territories that they vigorously defend.

Bony fish also show diversity in their breeding strategies. Many schooling fish are broadcast spawners that mate in large gatherings. Some gather only to reproduce. Most usually produce tremendous quantities of fertilized eggs to ensure the survival of the next generation.

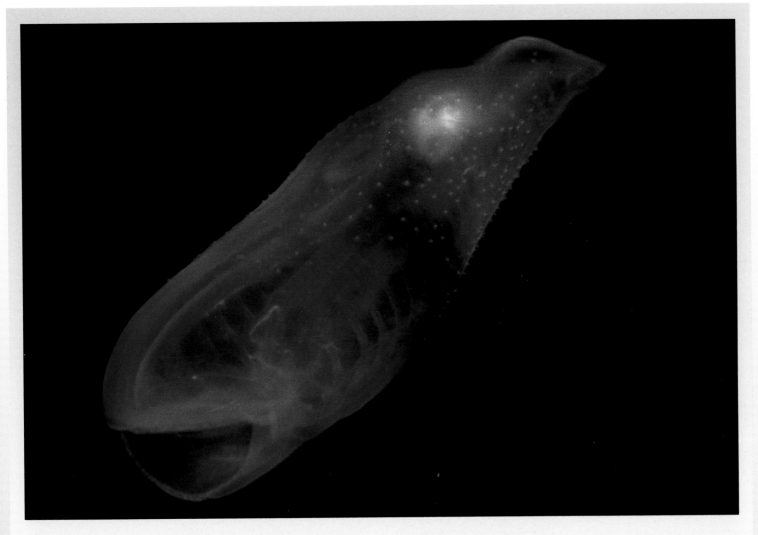

◄‼► Opposite page: *These bluebell tunicates look like clumps of juicy grapes. Adult tunicates remain attached to the seafloor by adhesive pads located on their heads. When feeding, each individual tunicate filters water through an incurrent and an excurrent pore. As adults, some tunicates are solitary, while others live a colonial life. Most solitary tunicates are less than six inches tall, whereas large colonies often cover several square feet. Tunicate coloration varies from vivid shades of purple, yellow, and iridescent blue to off-white or brown.*

◄‼► Above: *Free-living tunicates, known as salps, spend their adult lives drifting in the open sea. They feed by gliding through the water with mouths open, swallowing and filtering large quantities of water. Some varieties of salp live in colonies called salp chains. Often curled into attractive spiral patterns, salp chains look somewhat like interconnected, translucent Christmas tree ornaments. Most chains are only a few feet long, but they can measure up to 40 feet in length.*

❦ Opposite page: *A marine iguana peers across a rock in the Galápagos Islands. In the early morning, iguanas assume a flat basking posture on the coastal rocks and catch the tropical sun's warming rays; by midday, the sun has sufficiently warmed their bodies, and they are ready to enter the cool island waters to feed. Marine iguanas are excellent swimmers and divers; the large males easily penetrate the surf and dive 30 feet or more to graze on the algae found in the off-shore waters. When the iguanas finish feeding, they return to land. There, they bask in the afternoon sun to raise their body temperature in preparation for the cool nights.*

❦ Right: *Normally, iguanas' skin is drab black or brown with a few grey, green, or reddish splotches; during the mating season, however, the males become mottled with bright green, red, or orange spots. Males do not control females but may battle other males within a given territory to determine dominance. During courtship, the male elaborately circles the female while nodding his head. After mating, the females find sandy areas where they lay two to four eggs that will hatch in two to four months.*

❊ Above: *An olive sea snake swims through the shallow tropical waters of a coral reef off Australia. All of the approximately 55 species of sea snakes have an entirely aquatic lifestyle, and all are excellent swimmers. Some snakes can remain under water for several hours, and dives often last 15 minutes or more. All species of sea snakes reside in the warm, shallow regions of the Indo-Pacific, where they feed primarily on small fish.*

❊ Opposite page: *Far from hiding, many sea snakes advertise their presence with their brightly patterned skin, discouraging predators from coming within striking distance of their highly venomous fangs. The chemical composition of sea snake venoms varies from species to species. All, however, are fast-acting neurotoxins that quickly incapacitate prey. As air breathers in a marine environment, the snakes are at a disadvantage when hunting prey animals that have gills. Their potent venom assures the snakes a quick victory so they can avoid long, energy-consuming battles beneath the water.*

◄‖► Opposite page: *This loggerhead sea turtle, one of seven species of marine turtles, swims in the warm Bahamian waters. At about six feet and 1,000 pounds, another species, the giant leatherback turtle, is the largest living turtle species. The smallest species, the Kemp's ridley, is just over two feet long and weighs about 80 pounds. The remaining four species are the green turtle, hawksbill turtle, olive ridley, and Australian flatback turtle.*

◄‖► Above: *A hawksbill turtle shows off its distinctive beak, which it uses to graze sponges on coral reefs. The diet of sea turtles varies from species to species and covers the full spectrum. Some species are herbivorous, others carnivorous, and still others eat both plants and animals. One species, the green sea turtle, changes its diet as it ages. During the first year or so, it hunts small mollusks, crustaceans, echinoderms, and some fish. After that, it feeds primarily on marine plants.*

◄╫► Opposite page: *Sea turtles spend almost their entire lives in the water. When resting, green turtles, like this one, can stay submerged for five hours. Some green turtles can hibernate underwater for months. The adults of most species inhabit relatively shallow waters and occasionally venture out into deeper areas, but leatherbacks spend most of their time in the open waters. Some turtles migrate great distances, but particular migration habits can vary among different populations of the same species.*

◄╫► Above: *A female green turtle lays her eggs in a sandy nest. Sea turtles mate in the ocean, but the females lay their eggs on shore—usually during the night at high tide to ensure the nest is above the waterline.* Right: *Newly hatched green turtles head directly for the safety of the water. The young usually hatch at night and quickly scramble toward the sea. As many as 90 percent do not survive the first year, succumbing to predators of many kinds; however, those that do survive live to be at least 20, and perhaps 50, years old.*

◄◊► Above: *Fish occur in four basic shapes: fusiform, laterally compressed, dorsoventrally compressed, and attenuated. This blue jack is a good example of the fusiform variety. Fusiform species are more or less torpedo-shaped. Most have a slightly rounded head and a body that tapers toward the back. Typical of fusiform species, the blue jack is a swift swimmer. Barracuda, tuna, billfish, and many sharks are other examples of fusiform species that rank among the fastest fish in the ocean.*

◄◊► Opposite page: *A school of blue tangs grazes on algae. Like angelfish, butterfly fish, and spadefish, tangs have bodies that are laterally compressed; that is, the fish are much taller than they are wide. Fish with a laterally compressed body design are not particularly fast. They generally live in and around coral reefs and rocky areas. Their shape allows them access to the numerous nooks and crevices in their habitats, where they find food and safety from larger predators.*

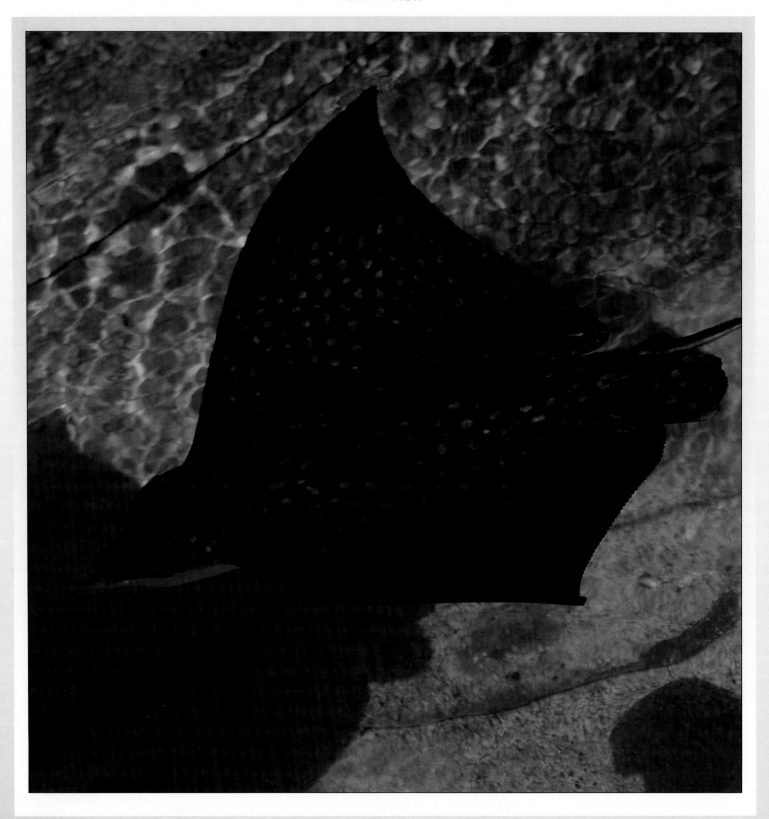

•‖• Opposite page: *A spotted eagle ray cruises over a stingray as it lies still on the sandy bottom. These animals bene-fit from their dorsoventrally com-pressed body shape both when moving and when stationary. Their wide, flat bodies give them grace, speed, and maneuverability as they glide through the water. Lying flat on the seafloor, they remain hidden from both prey and predator. Some rays are pelagic, open-ocean swimmers, and others spend nearly all their time resting on the seafloor. Skates, angel sharks, halibut, turbot, sole, and sanddabs are other dorsoventrally compressed fish.*

•‖• Right: *A primeval moray eel slithers along the ocean bottom. Members of the eel family have snakelike, or attenuated, bodies. These species maneuver easily in the restricted confines of rocky areas and coral reefs, making them formidable predators. Other groups noted for their attenuated bodies include the various species of pipefish, needlefish, and trumpetfish. While these fish are not especially fast long-range swimmers, many of them can make very quick and accurate strikes at their prey.*

◦❦◦ Right: *The sea horse may well be the most improbable of fish. It obviously does not match any of the four typical fish body shapes. With its armored coat, miniature fin, prehensile tail, and upright posture, it serves as perhaps the best testament to the incredible diversity of the fish family. When not maneuvering with its miniature fin, the sea horse anchors itself down with its curled tail. A slow, awkward swimmer, it relies primarily on camouflage and a heavily plated body to counter predators.*

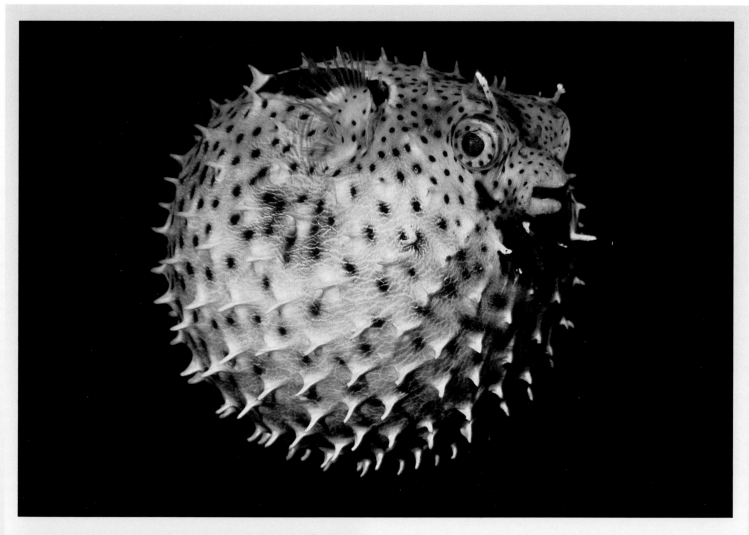

Above and left: *Another exception to the four basic body shapes comes in the form of the prickly puffer fish. In striking contrast to the inflexible bodies of sea horses, puffer fish bodies are inflatable. When threatened, a puffer fish will swallow water, enlarging itself into a round, spiny ball that looks more like a giant pincushion than a fish. An inflated puffer fish must appear an unappealing meal to all but the hungriest predators. When not inflated (left), this fish is half the size and its spines lay relatively flat.*

◄▮► Above: *Bony fish employ a variety of reproductive strategies to help ensure the continuation of their species. This male sergeant major, in charge of guarding its nest of fertilized purple eggs as they lie on the seafloor, will attack or remove any intruders in his nesting territory. It may take him several attempts to move this sea star, but his efforts will help ensure the survival of the next generation. Other nest builders include damselfish, clownfish, gobies, and triggerfish.*

◄▮► Opposite page: *A number of reproductive strategies involve protecting the developing embryos by means of a special brooding area on the adult fish itself. The jawfish is a mouth brooder. The male carries the fertilized eggs within its mouth until the young reach the hatching stage. During this period, which will last for many weeks, the adult male does not consume any food. Once hatched, the juveniles may stay close to their father for an additional month, moving in and out of the mouth at will, thereby obtaining protection until they venture out on their own.*

◀▮▶ Above: *The parrotfish's jaws each contain a number of individual teeth fused together to form its distinctive beak. These medium-size fish of the Atlantic and Pacific coral reefs are grazers. They feed primarily on the low growth of algae found on coral rock. They use their strong beaklike teeth to scrape the algal growth and, at times, the coral itself. Special molarlike teeth on the floor and roof of the throat then grind the algae and coral as it is swallowed.*

◀▮▶ Opposite page: *Bottom-feeding fish use many strategies to capture their food. These yellow goatfish (Mulloidichthys martinicus), found in the tropical waters of the Caribbean, rely on the whiskerlike barbels under their chins to locate their prey. The fish root in the sandy bottom with the chemical-sensitive barbels, detecting buried creatures such as worms and small crustaceans. When they locate any prey, they will quickly excavate it with their small, downward-tilting mouths and then devour it.*

◀▮▶ Opposite page: *A tarpon gorges itself on silversides. Many prey fish, such as mackerel and these silversides, rely on schooling as a means of protection. The idea at work here is safety in numbers. If a predator attacks a school, the chances of survival for any single fish are higher when the school is larger. In other words, each individual hides within the school, hoping that it will not be one of the victims. Predators may also be confused or distracted by all the motion within a school, making it difficult for them to target and pursue any one fish.*

◀▮▶ Above: *Huge schools of anchovies are a common sight in every ocean. Of the more than 20,000 known fish species, about 80 percent school as juveniles and 20 percent school as adults. Despite a great deal of study, questions remain as to why fish school. Some benefits, however, seem certain. In addition to providing some measure of safety from predators, schooling enhances the chances that males and females will find each other and mate; some species school specifically for that purpose. Schooling also can make swimming easier by reducing water resistance.*

◀‖▸ Above: *Schooling behavior is not limited to the smaller species of baitfish. Here, a school of barracuda prowls the Celebes Sea off Borneo. The smaller species of barracuda, in particular, often congregate in schools. Barracuda can accelerate from a dead stop to full speed almost instantly, making them extremely capable hunters. Whole schools of barracuda will relentlessly hound small baitfish, such as anchovies and sardines. When pursuing larger prey, barracuda will sweep through a school snapping their jaws to cut prey in half and then turn back to collect the pieces.*

◀‖▸ Opposite page: *Large, fast-swimming jacks school, albeit in smaller numbers, as they search out prey. Predators such as jacks and tuna counter the schooling tactics of their prey fish by forming schools of their own. Working together, the predators can break the school of prey into smaller, less organized groups. By forcing their prey into smaller groups, they create easier targets. Their organized attacks will also confuse some individuals and cause them to separate from the safety of the school.*

◄▐▐► Opposite page: *Most small fish would be eaten by an anemone this size, but clownfish (also called anemonefish) can spend hours within an anemone's dangerous stinging tentacles. These species have developed a symbiotic relationship that benefits both animals. Predators cannot pursue the clownfish without almost certainly suffering fatal stings; also, when the anemone captures food, the clownfish may be able to take some of the scraps for itself. The anemone benefits because its guest will drive away potential predators and sometimes may provide the anemone with leftovers from its meals. The fish also helps by removing parasites and waste products from the anemone.*

◄▐▐► Right: *Exactly how the clownfish avoids being harmed by the anemone is not altogether known. The fish initiates a relationship with an anemone by making very brief contacts with the anemone's tentacles. Although the fish is stung, the contacts continue and intensify. Eventually, the fish is able to rest in the anemone's tentacles unharmed. Possibly, the contact with the anemone stimulates the fish to produce a heavy layer of protective mucus on its body. If a fish breaks off contact with an anemone for an extended period of time, it must reacclimate itself to that anemone if it returns.*

◀ǁ▶ Above: *Perhaps some of the most remarkable interactions among marine animals are cleaning relationships. Here, a grouper is cleaned by a few small gobies. The gobies will rid the grouper of parasites, dead tissue, loose scales, mucus, and bacteria—and get themselves a meal in the process. In Caribbean reef communities, various species will congregate in large numbers as gobies and blennies diligently clean groupers, coneys, rock hinds, and any number of other fish. Most cleaners do not rely solely on their cleaning activity to gather food.*

◀ǁ▶ Opposite page: *A creole wrasse signals that it is ready to be cleaned by hovering in the water at a slight angle, mouth open, head down, tail up. Such posturing lets cleaners know that their services are wanted and that they face no danger from the larger fish. Once the wrasse has had enough, it will again perform a distinctive series of moves to let the cleaners know they should depart. Other fish species perform different but equally recognizable posturing behaviors.*

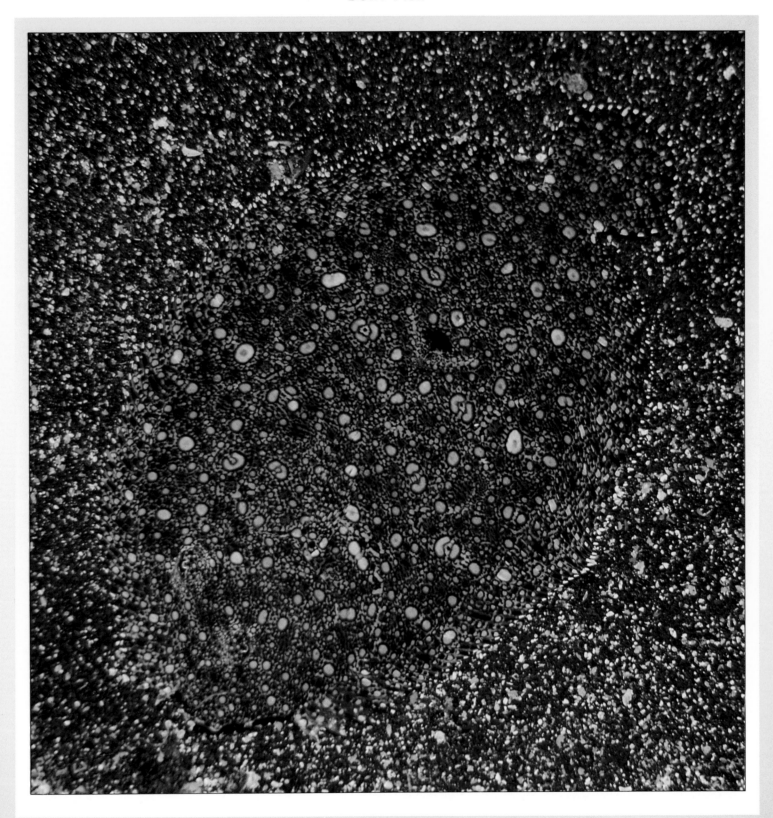

❦ Opposite page: *This spotted flounder in the waters off Big Island, Hawaii, is one of the most versatile and effective camouflage artists found in nature. Its flattened body shape and remarkable ability to change colors allow it to blend almost perfectly with its environment. These fish are able to alter their coloration to match almost any color or pattern and thereby become virtually invisible to predators and prey. Other types of flatfish, such as turbot, halibut, and sole, have comparable camouflage abilities.*

❦ Above: *A sole disguises itself amidst the shells and rocks on the ocean floor. Specialized pigment cells called chromatophores allow these fish to execute their rapid and dramatic color changes. These irregularly branched cells each contain a particular color pigment, which is normally concentrated at the center of the cell. When the fish's nervous system directs the pigment to disperse throughout the cell, the color becomes prominent. By turning individual chromatophores on and off, this sole can create a pattern that matches not just the color but also the size of the objects that surround it.*

◀▶ Above: *The peacock flounder* (Bothus lunatus) *and several other flatfish have developed an interesting trait in response to their bottom-dwelling lifestyle. The free-swimming larva of these species have one eye on either side of the head. As the fish matures and prepares for a life of lying on its side on the seafloor, one eye gradually migrates to the other side of the head so that the fish can lie on the bottom and still retain its full range of vision.*

◀▶ Opposite page: *Two lizardfish demonstrate this species' remarkable knack for camouflage. One of the lizardfish has already altered its color to blend in perfectly with the background. Chances are, the other one will soon follow suit. Lizardfish, flatfish, and trumpetfish frequently change their coloration; many other species rarely or never change color. Some species use color changes for reasons other than camouflage. Territorial fish, for example, sometimes flash warning colors at intruders in their domain, while some other species change color as a way to attract or signal a mate.*

◄‖► Above: *Some fish alter their color to match their surroundings from moment to moment, while other fish live where their coloration and shape constantly match their surroundings. This leafy sea dragon will blend right into the kelp plants on which it feeds, thanks to the frilly, kelp-colored appendages that decorate its body. Relatives of the sea horse and pipefish, sea dragons are slow swimmers, but their marvelous camouflage allows them to hide safely in the low-lying kelp.*

◄‖► Opposite page: *The sargassum fish lives exclusively in the floating mats of sargassum weed that serve as a habitat for many species in the Atlantic Ocean's Sargasso Sea, southeast of Bermuda. Its ornate coloration and frilly appendages keep the small fish hidden from predators amid the heavy tangle of weeds, but this species uses its camouflage to hunt as well as to hide. The small fish and crustaceans that it feeds on will unwarily move within striking distance as they graze on the floating mat of weeds.*

◄❙▶ Opposite page: *A trumpetfish hovers motionless alongside a soft gorgonian coral. Shape, color, and posture all work together to help it evade detection. Thus disguised, it will wait patiently and safely for a smaller fish to wander past. With amazing speed, the trumpetfish can then dart forward and open its surprisingly large mouth. When still some distance from its prey, the trumpetfish will sharply suck in a stream of water and pull its meal in along with it.*

◄❙▶ Above: *A relative of the sea horse, this ornate ghost pipefish has a rather loud, noticeable color pattern, but its elongated body and spiky appendages will help to make it considerably more discreet. All 150 known species of pipefish live in shallow tropical and subtropical waters. Poor swimmers, they use their prehensile tails to anchor themselves to plants, corals, and other structures while they wait for tiny crustaceans and other prey to float by.*

◄◗► Opposite page: *A stonefish easily blends into the background near the Seychelles island group in the Indian Ocean. In a tactic called aggressive mimicry, the fish relies on its irregular body shape, knobby surface, and mottled coloration to become invisible to its prey. It also has an added defense against any potential predators—highly venomous spines that line its back and sides. The extremely strong toxin is fatal to humans and serves as an effective back-up to the camouflage defense.*

◄◗► Above: *A scorpionfish uses the same kind of mottled coloration and irregular body shape to blend seamlessly into a rocky outcrop. Like the stonefish, the scorpionfish also relies on a very potent toxin in its dorsal spines to ward off any enemies. This bottom-dwelling reef fish is a poor swimmer; it is neither fast nor agile, but its powerful toxin and its excellent mimicry of its environment leave it safe from virtually all predators.*

◀▮▶ Above: *The bright, arresting stripes and long, frilly fins make this lionfish easy to identify. Like stonefish and scorpionfish, lionfish are slow-swimming reef dwellers that carry a deadly poison in their dorsal spines. In marked contrast to the two other species, lionfish have a color pattern and body shape that make them stand out clearly against almost any background. Predators identify the bright warning coloration as a source of danger and learn to keep their distance.*

◀▮▶ Opposite page: *The unusual dark and light patches of these butterfly fish break up the outline of the individual fish and make them hard to identify at a glance. Also, each fish has a false eyespot on its tail and a black stripe hiding its real eye, which can create the illusion that the fish is facing the other way. Predators often direct a strike at their prey's head, assuming that the prey will flee in that direction. When attacking one of these butterfly fish, a predator might aim at the tail by mistake. This slight misdirection could give the butterfly fish just enough of an advantage to escape.*

The Shark Family

SHARKS, RAYS, AND SKATES make up a scientific grouping of fish called elasmobranchs. All elasmobranchs have skeletons of cartilage; most fish have skeletons of bone. As a group, sharks, rays, and skates have lived on Earth for over 450 million years. Their fossil record dates back more than twice as long as that of the dinosaurs. Because cartilage precedes bone in the skeletal development of animals, many scientists thought that bony fish evolved from the cartilaginous fish. This belief contributed to the misconception that sharks, rays, and skates are primitive, simple animals that lack intelligence. Recent discoveries, though, have supported modern theories that the two groups evolved independent of each other.

SHARKS

Few images convey the terror felt when we see a shark swimming toward its prey with mouth agape and razor-sharp teeth fully exposed. Many people see sharks as mindless eating machines, voracious hunters, and indiscriminate feeders that become frenzied when they sense a single drop of blood. Scientists and divers who have worked with sharks know this idea is flawed. In reality, sharks are as misunderstood as they are feared.

Perhaps the first thing to learn about sharks is that there is no single animal called the shark. Worldwide there are more than 350 species. While the species share many common characteristics, they differ widely in their size, where they live, what they prefer to eat, and how they fit into nature's overall design.

The typical shark has a slender, graceful body that is slightly thicker in the middle and tapers at both ends. The snout, or nose, ranges from pointed to round to squared off. Typical sharks have a pair of dorsal fins located along the top of their back and a sickle-shaped tail whose upper lobe is larger than the lower lobe. The mouth of most species sits on the underside of the head. This general description applies to a wide range of species such as gray reef sharks, blue sharks, whitetips, sand tigers, bull sharks, lemon sharks, tiger sharks, and many more.

However, there are significant variations in body design in other species. The bodies of angel sharks and wobbegongs are greatly flattened. Mako sharks, porbeagles, salmon sharks, and great whites have nearly symmetrical tails, with the upper and lower lobes almost the

Opposite page: *A whitetip reef shark cruises in Hawaiian waters.* This page: *This sluggish epaulette shark from the Indian Ocean will spend much of its time on the seafloor.*

🦈 Above: *This deep-water sevengill shark is one of the very few species that has more than five pairs of gill slits.*

Like all fish, sharks have special nerves that run laterally along the sides of their bodies. These nerves, collectively referred to as the lateral line, help them sense vibrations in the water around them. Sharks do have inner ears, but the extent to which they rely upon hearing remains a point in question. Certainly hearing does play a part in the way sharks analyze their surroundings, but it is difficult for scientists to determine whether the ears or the lateral line do more to detect and analyze sound.

Vision varies considerably from species to species. Most scientists believe sharks use sight to some degree to help analyze stimuli from a distance. Up close, within a few feet of their prey, sight is not as important as some other senses.

One of the most remarkable senses in sharks is their ability to detect electrical fields. They do this with special organs called the ampullae of Lorenzini, which are small, gel-filled pits in the snout and other parts of the body. All living organisms create electrical fields around themselves, and this ability helps many sharks locate food sources hidden in crevices or buried in sand.

identical shape and size. And the mouths of filter-feeding whale sharks are at the front of the head rather than underneath.

Sharks lack the internal organ called a swim bladder that assists in buoyancy control in most bony fish. Still they are very efficient swimmers. They gain their forward thrust with the back and forth movement of their long, powerful tails, and they use the pectoral fins along their sides to control turns. The cartilaginous skeleton of sharks also aids in swimming. Cartilage is lighter and more pliable than bone. The increased flexibility contributes to the sharks' graceful, sinuous swimming motion.

Well-developed senses and many other adaptations have made sharks one of the dominant predators in many marine habitats for millions of years. They use a combination of senses to acquire food and analyze surroundings.

Most of a shark's olfactory, or smell, receptors are small sacs in its nostrils, and they are very well developed. Laboratory experiments show that some sharks can detect one part of blood in one million parts of water.

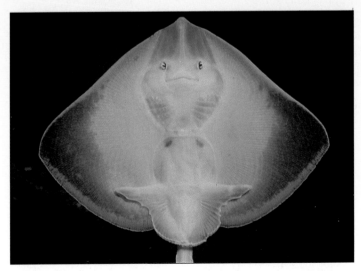

🦈 Above: *Unlike any of its shark relatives, this common skate* (Raja batis) *clearly has its gill slits set on the underside of its body.*

Different species have evolved specific types of teeth that help them catch their prey. Some sharks have long, spiked teeth that allow them to snag small, fast-moving fish. Others have serrated triangular teeth that cut chunks of flesh from larger prey. Still others have broad, flat teeth for crushing the outer coverings of shellfish. Sharks also have many rows of teeth within their mouths, when a front tooth is lost during the struggle to capture prey, a new one moves up from behind to take its place.

SKATES AND RAYS

Skates and rays, close relatives of sharks, are represented by about 470 different species. Skates look very similar to rays and to some bottom-dwelling sharks. A simple way to distinguish among sharks, rays, and

Above: *A stunning blue-spotted stingray rests on the rocky floor of the Red Sea.*

skates is to look at the gills and tail. A shark's gill slits are always on the side of the head, while the gills of rays and skates are always on the underside of the body. Also, rays have thin, whiplike tails, while sharks and skates have fleshier, more prominent tails.

Rays thrive throughout tropical and temperate seas. Species such as manta rays and mobula rays are pelagic creatures, meaning they prefer to stay in the waters of the open sea. Others, such as stingrays, bat rays, and spotted eagle rays, frequent reef communities. Manta rays and mobula rays rarely, if ever, approach the seafloor, while many species of stingrays live their entire lives near the bottom.

🦈 Above: *Angel sharks are a bottom-dwelling temperate species. Although they are lithe, graceful swimmers, they spend most of their time lying on the seafloor. Angel sharks, like many other bottom dwellers, rely heavily on camouflage to avoid potential predators and capture prey. With its coloration, this angel shark can rest virtually unnoticed on the sandy bottom, awaiting its prey—usually crustaceans, echinoderms, and small fish. So effective is its camouflage that prey will often swim unaware within inches of an angel shark's mouth.*

🦈 Opposite page: *Angel sharks have extremely flattened bodies—many divers mistake them for rays—and they blend seamlessly with their environment. As with most bottom dwellers, they are comparatively small and quite docile. This makes them seem approachable to many people. However, divers still need to exercise caution and common sense. By foolishly trying to ride or handle these seemingly tame creatures, divers and snorkelers have been known to provoke brutal attacks.*

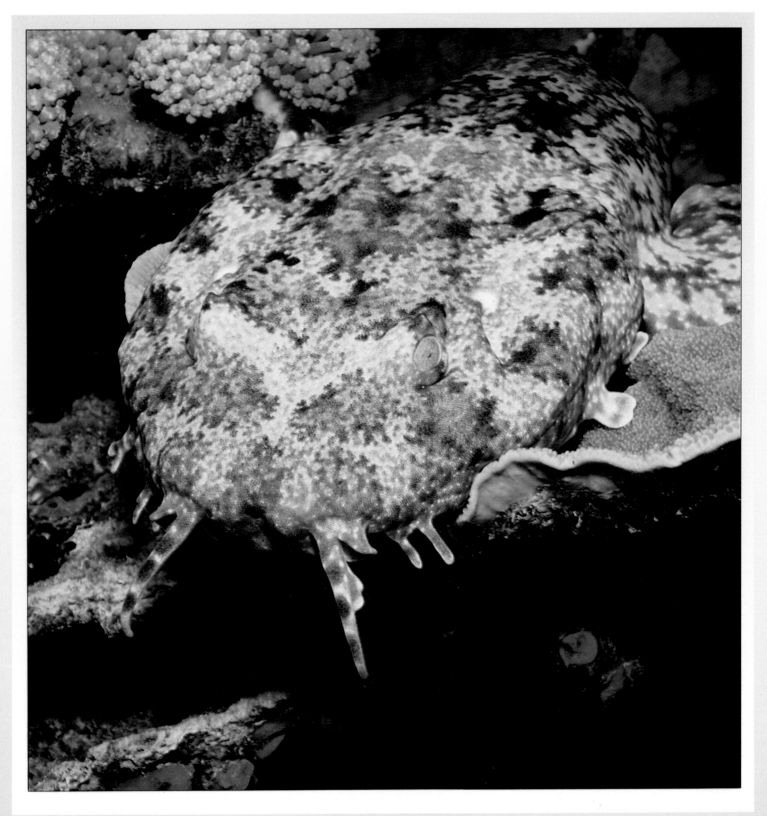

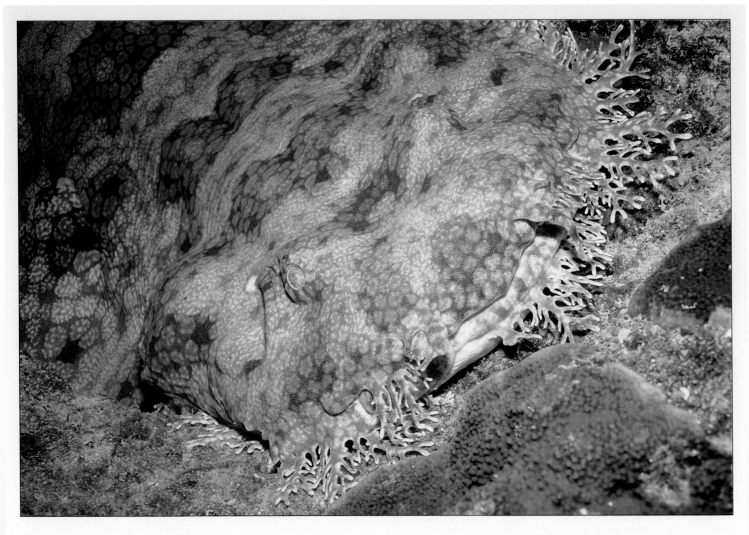

Opposite page: *Like angel sharks, wobbegongs have greatly flattened bodies. This tasselled wobbegong is well camouflaged: the blotches, spots, and lines on its exterior and the unique lobes of skin running along the sides of its head help it to blend into its surroundings. Indeed, wobbegongs use camouflage as much as strength to feed on a variety of small prey—mostly bottom fish, crabs, and lobsters—which they grasp with their sharp, fanglike teeth.*

Above: *Wobbegongs, along with banded catsharks, nurse sharks, and whitetip reef sharks, are among the most common bottom dwellers in tropical reef communities. This ornate wobbegong (Orectolobus ornatus) sleeps along a rocky reef. Wobbegongs spend much of their time resting on the ocean floor, sometimes moving only to hunt for food. Research has shown that over their whole lifetime many of these sharks do not venture more than a few miles from where they were born.*

🦈 Right: *A baby swell shark in its egg case is less than two inches long. Females deposit these leathery cases on the seafloor, often in heavy vegetation, to hold and protect the developing eggs. Not all sharks produce egg cases; in some species the females lay eggs, and in others the eggs develop and hatch inside the mother's body. Still other species give birth to live young.*

🦈 Above: *A swell shark in the waters off the coast of California. When disturbed, these three-foot sharks can inflate with water like a puffer fish. Swell sharks, like angel, horn, and nurse sharks, can breathe without swimming because they are able to pump water over their gills. This makes it possible for them to rest on the ocean floor for extended periods of time. Many other sharks, such as blue sharks and mako sharks, must swim constantly in order to create a flow of oxygen-rich water over their gills.*

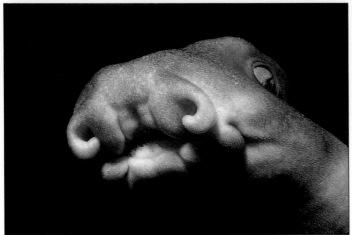

Above: *Upon first sighting, some shark species, such as the horn shark shown here, look more like catfish than sharks. The horn shark is identifiable by the sharp spines in its dorsal fins—spines that could lodge in the throat or mouth of a predator. Found in the kelp forest communities of western North America, horn sharks reach a maximum size of about three feet and are very tame creatures. In fact, the horn shark is one of the few sharks that can be kept in captivity. Left: Hardly the image of teeth and power usually associated with sharks, the horn shark has tri-cusped frontal teeth, the central cusp being the largest, and its lateral teeth are modified into flat molars.*

Left: *A nurse shark moves after a school of porkfish and grunts. At this close range, the shark will rely on the electroreceptors in its snout to home in on its prey. The sense organs that allow sharks to detect the electrical fields surrounding all living organisms are called the ampullae of Lorenzini. These small, gel-filled pits in the snout and other parts of the shark's body let it detect electrical fields far more faint than any other known animal can distinguish. In addition to helping the shark locate its prey, the electroreceptors may very well serve a navigational function.*

Right: *Known in the United States as the sand tiger shark and in Australia as the grey nurse shark, it is the South African name—raggedtooth shark—that perhaps best captures the apparent power and ferocity of this animal. These creatures generally look much fiercer than they really are. The sand tiger's long, spikelike teeth make it perfectly suited for capturing and eating its preferred diet of small fish. Sand tiger sharks are found in warm and temperate waters throughout the world.*

 Above: *In contrast to the sand tiger shark, the leopard shark* (Stegostoma fasciatum) *has teeth that are short and broad for crushing hard-shelled prey, such as crabs and lobsters. The leopard shark's striking coloration makes it easily recognizable. Leopard sharks are found from the Oregon coast to the Gulf of California and are common in inshore waters from spring to midsummer. Known nursery areas for this species include the San Francisco, Tomales, and Bodega bays. Leopard shark pups measure seven to eight inches in length, and litters range from seven to thirty pups.*

🦈 Above: *Scalloped hammerheads demonstrate a rather unusual grouping or schooling behavior. As a rule, most sharks never form large groups except when attracted by bait or when mating. Scalloped hammerheads, however, often gather by the hundreds in many parts of the tropical Pacific. Individuals within the group frequently show distinctive swimming patterns that might serve as some form of communication among the animals. Despite more than a decade of study, scientists still have not learned the reason that these sharks school.*

🦈 Opposite page: *Hammerheads are relatively modern sharks, having evolved some 120 million years ago. The most striking feature of the nine different species in this group is, of course, the hammer-shaped head. This unusual design may serve two important functions. First, the head's shape provides added lift at the front end of the body, similar to the way wings provide lift to airplanes. In addition, a wider distribution of sensory organs may enable hammerheads to locate prey and predators more quickly. Inset: A close-up of a young hammerhead's mouth.*

🦈 Below: *The mako shark is an open-ocean species in the same family as the great white. Most fish are cold-blooded, meaning they cannot internally control their body temperature. The warm-bodied mako is able to maintain a temperature several degrees above that of its environment, which allows its muscles to work more efficiently. This seemingly slight evolutionary advantage makes the mako one of the fastest animals in the sea, allowing it to successfully pursue its usual diet of fast-swimming tuna and swordfish.*

🦈 Opposite page and above: *Great white sharks are among the most awe-inspiring—and most feared—creatures in the sea. Their size and power make them a threat to almost any animal in their domain. Specimens have been reported to reach more than 30 feet, and at that size a great white would easily weigh 10,000 pounds or more. Adults typically range from 12 to 16 feet in length and weigh from 1,800 to 3,500 pounds; newborns are about 3 to 4 feet long. Great white sharks have tri-angular teeth whose serrated edges look much like the blade of a saw. Individual teeth can be over two inches long.*

🦈 Above: *Though great whites live in all oceans, people most often encounter them in the temperate shallow waters of continental shelves. Great whites are not the indiscriminate feeders that people commonly think they are. During the first few years of life, they feed upon a variety of rays, flatfish, and other animals that live in sandy areas or in nearby reef communities. As they become larger, their diet shifts toward marine mammals. Adult great whites hunt sea lions, seals, and dolphins. They also sometimes feed on other sharks and on salmon, sturgeon, hake, tuna, and other large fish.*

🦈 Opposite page: *Like most animals, sharks do not rely on a single adaptation to capture prey and survive in an ever-changing environment. Sharks are well known for their impressive array of deadly teeth, but they have other features that give them an edge against prey. This great white demonstrates its protrusible jaws. The jaws are loosely attached to the cranium, and special ligaments extend them forward as the shark bites, increasing its reach and putting the teeth in a better position to secure large prey.*

Above: *A gray reef shark* (Carcharhinus amblyrhynchos) *cruises the waters off the island of Hawaii. Gray reef sharks belong to one of the larger shark families, the requiem sharks. The group has an extremely wide distribution, ranging throughout virtually all temperate and tropical seas. Requiem sharks are voracious predators and are well documented for their feeding frenzies. This group is regarded as the most economically important shark family; many species are utilized for food, oil, and leather, among other things.*

Opposite page: *Gray reef sharks are typical of most reef sharks. While these sharks do not live in schools, they tend to congregate in reef communities. This species is capable of rapid bursts of speed and has excellent maneuverability, a necessary feature of predators that operate in the tight confines of the reef. Gray reef sharks are one of the few species that are fiercely territorial. They are well known for their exaggerated posturing, radically hunching their back and tucking their pectoral fins before they strike. These threat displays do not precede all attacks by gray reef sharks, but once the sharks display, they almost always attack.*

🦈 Above: *A lone whitetip reef shark silently cruises the shallow waters off the coast of Hawaii. Whitetips may be one of the most common bottom-dwelling sharks found in tropical reef communities. Like many large, bottom-dwelling reef species, whitetips feed primarily on mollusks, echinoderms, and small fish. Whitetips belong to the small group of sharks that bear live young, an unusual reproductive characteristic among fish. This advanced form of reproduction also appears in other requiem sharks, smooth dogfish sharks, and hammerheads.*

🦈 Right: *An oceanic whitetip shark in the Pacific is accompanied by pilot fish. These fish swim near the shark's mouth waiting for scraps left over from the shark's meals. Oceanic whitetips hunt squid, tuna, marlin, and barracuda, but they are not above feeding on the carcasses of dead whales if they are available. Oceanic whitetip sharks are quite common in open tropical seas; they rarely if ever venture into waters less than 600 feet deep.*

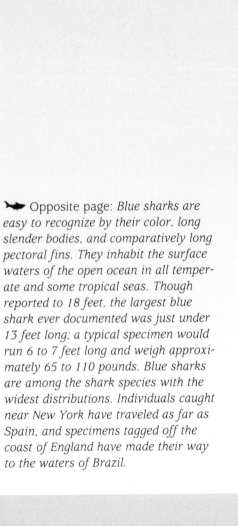

Right: *Named for their iridescent color, blue sharks are among the most beautiful of all shark species. Their magnificent color is most obvious on sunny days when the sharks swim near the surface. Like many open-ocean predators, blue sharks are counter-shaded: They have a dark top side and a lightly colored underside. Counter-shading works because sunlight does not penetrate deep water. Thus, when viewed from above, the dark upper half of the animal blends in with the dark background of deep water; when viewed from below, the shark's light underbelly blends into the light-colored surface water.*

Opposite page: *Blue sharks are easy to recognize by their color, long slender bodies, and comparatively long pectoral fins. They inhabit the surface waters of the open ocean in all temperate and some tropical seas. Though reported to 18 feet, the largest blue shark ever documented was just under 13 feet long; a typical specimen would run 6 to 7 feet long and weigh approximately 65 to 110 pounds. Blue sharks are among the shark species with the widest distributions. Individuals caught near New York have traveled as far as Spain, and specimens tagged off the coast of England have made their way to the waters of Brazil.*

Above: *Lights from a boat illuminate a blue shark as it feeds on squid in the dark night waters. Blues are commonly drawn to the huge groups of squid that gather to mate in open waters. They also prey regularly on small schooling fish, such as anchovies and herring. Like other open-ocean species, blue sharks can grow as much as 10 to 15 inches per year when food is abundant.* Left: *The nictitating membrane begins to close over a blue shark's eye as it moves in to feed. The tough membrane protects the eye from damage that could be inflicted by prey thrashing wildly in the shark's jaws.*

 *Left: A medium-size blue shark helps a diver test a "shark-proof" diving suit. Known as the Neptunic, the suit is composed of thousands of interlocking steel rings, much like chain mail armor. The basic idea behind the suit is simple—the circular rings are too small for the shark's teeth to penetrate them—and it has been found to be quite effective. The enormous pressure of a bite from a shark's powerful jaws can still crush bones or damage internal organs, but by countering the shark's primary weapon, its teeth, the suit can still save a diver's life.*

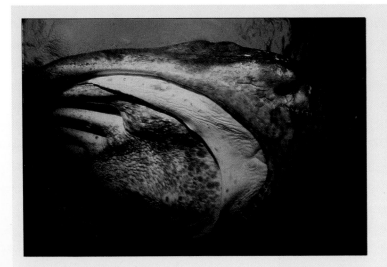

🦈 Above: *A basking shark displays its huge mouth as it strains plankton from the water through its gills. Basking sharks live in the northern temperate and subpolar oceans, where plankton is readily available, but they occasionally stray south into warmer areas. Although they are the second-largest fish in the world, basking sharks are not well studied. They seem to have definite migratory patterns related to the plankton supply and to their mating cycle, but scientists know little else about their behavior. They all but disappear during the winter months, leading some to suggest that they might hibernate; it's also possible that they move to very deep waters during this time, which would account for the lack of sightings.*

🦈 Right: *The whale shark easily lays claim to the title of biggest fish in the sea. Individuals can reach 40 feet in length (at least 10 feet longer than the basking shark) and weigh an astounding 30,000 pounds or more. The mouth of a large whale shark can be ten feet across and contain over 5,000 teeth. Whale sharks restrict their movements to the tropical regions, where they strain the water for enormous quantities of small crustaceans and some fish under about six pounds.*

Opposite page: *Whale sharks are one of the few species that experienced divers actually feel comfortable in the water with. Despite their imposing size, they pose no threat at all to humans. They are gentle and calm and quite receptive to swimmers that approach them. It's not uncommon for divers to touch them or even grab onto a fin and hitch a ride as the huge creatures swim.*

Above: *The most obvious way to identify a whale shark is by its size, but several other features set it apart from most sharks. The wide mouth appears at the front of the head rather than on the underside of the body, and the snout is very definitely squared off rather than tapered. The distinctive color markings are atypical also. They are often a reddish or greenish gray on top with a striking array of yellow or white spots and stripes; underneath they are most often a yellowish white.*

➤ Above: *A sand skate lies quietly on the seafloor. Most skates spend the daylight hours resting in just such a position. At night, they rise from the bottom and seek out their evening's meal, most often small mollusks and crustaceans. Currently, over 400 species of rays and skates are known. Like sharks, they have a skeleton of cartilage, they lack a swim bladder, and they are covered with rough skin composed of dermal denticles, which are very similar in structure to teeth.*

➤ Opposite page: *The distinctive-looking sawfish was given its common name for obvious reasons. This large skate uses its barbed, elongated snout to probe the bottom for prey; it will also swim through schools of small fish rapidly whipping its head from side to side to stun and injure prey, and then wheel around to collect the wounded animals. Its cousin the sawshark has evolved a similar snout and uses it in much the same way; the sawshark has the added advantage of a pair of sensitive barbels on its saw that aid in locating buried prey.*

🦈 Above: *Surely the most impressive of all the rays is the manta ray. People once viewed them as dangerous monsters of the deep, but they are actually very docile and tolerant. This diver is quite fortunate; people rarely get such a close-up look at mantas. The huge, flattened pectoral fins provide considerable thrust and can move independently, making mantas excellent swimmers. The rays can easily outdistance humans with only the slightest movement of their fins.*

🦈 Opposite page: *Manta rays can grow to 20 feet across, and at that size they would weigh in the neighborhood of 3,000 pounds, making them easily the largest member of the ray family. Despite their bulk, they move with remarkable grace and elegance. They are powerful creatures, too. They can generate enough thrust to break through the surface of the water and soar several feet through the air. The reason for this behavior is not understood, but it may be a method of removing external parasites.*

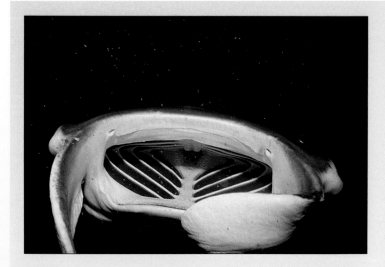

➤ Above: *A manta ray filter feeds in the nighttime waters of the Pacific near the Hawaiian Islands. Like many large ocean creatures, the manta ray feeds on tiny plankton that it strains from the water. The appendages on either side of its mouth are cephalic fins, a distinguishing feature of mantas. The manta uses them to guide quantities of plankton-rich water into its mouth as a way to feed more efficiently. The cephalic fins look somewhat like horns and are probably what earned the manta the nickname of "devilfish."*

➤ Opposite page: *A thin, whiplike tail with no dorsal fin— one of the distinguishing features of manta rays—trails behind this specimen as it glides effortlessly over the sandy bottom. Mantas are pelagic creatures, meaning they spend their lives endlessly cruising the waters of the open sea. It's unusual to come across one in such shallow waters, but occasionally they will venture closer to shore.*

➤ Opposite page: *A bar jack disturbs a southern stingray as it rests in the sand. These rays are common in the Caribbean and in the warmer waters of the Atlantic Ocean. They are equipped with a sharp spine in their tail that they use primarily as a defensive weapon. The spines are composed of denticles, a toothlike material that grows in successive rows; when the tip of the spine is broken or pulled off, new denticles move up to replace it in much the same way that sharks replace their teeth.*

➤ Above: *Stingrays frequently bury themselves in the sand with only their eyes protruding as a way to hide from predators as well as prey. The technique serves the rays well but can cause problems for swimmers and snorkelers. Many people have inadvertently disturbed or stepped on a buried ray only to be struck with the tail spine. Such a wound can be quite painful. In addition to the spines, stingrays have venom glands on either side of their tails; the venom will greatly inflame the wound and has even resulted in death in humans.*

Opposite page: *Superb swimmers, eagle rays often haunt the seaward edge of tropical reefs. From there, they can enter the reef zone at will to hunt crustaceans and mollusks or quickly retreat to the open sea where they can outswim any would-be predators. It is quite common to find these active creatures cruising in large groups like this. Eagle rays frequently reach a six-foot width and can weigh over 500 pounds.*

Above: *The spotted eagle ray is decorated with a stunning pattern on its top side. The eagle ray has an unusually long tail; it can be over three times the length of the ray's body. Like the stingray, it has a series of spines in the tail, but the spines are at the base of the tail, close to the body. In such a position, it is difficult for the eagle ray to whip the tail around and strike with the spines, so their use in defense is probably rather limited.*

➤ Above: *Most rays are excellent swimmers, but of course there are exceptions. The electric ray shown here is a sluggish bottom dweller from the Red Sea. These torpedo rays, as they are also known, have unusually fleshy tails for rays. They also have large patches of specialized muscle tissue on their pectoral fins that produce electricity. These electrical organs comprise about one-sixth of the fish's body weight. The electrical intensity varies from species to species, but the shocks are quite effective in battling predators and capturing prey. Creating the electricity requires much energy, and the rays use it only sparingly.*

➤ Opposite page: *Like the manta ray, the mobula ray is another graceful cruiser of the open seas. It is also one of the larger rays and has the prominent cephalic fins clearly visible on either side of its head. This specimen has a remora fish attached to its underside. Remoras spend their lives keeping company with large pelagic animals such as rays, sharks, turtles, and billfish, feeding on scraps and leftovers from their companions' meals.*

Opposite page: *A bat ray (Myliobatis californica) rests in the Pacific waters off San Clemente Island. Bat rays are medium-size predators that frequent reef communities. Their unusually thick bodies and blunt, boxy heads make them rather easy to identify. They are much less aerodynamic than the typical ray, but they are still able to maneuver deftly through the rocky confines of their reef habitats.*

Above: *This southern fiddler ray (Tygonorhina guanerius) is parked off the southern coast of Australia. They grow to about three feet long and have a distinctive marking pattern on their top sides. This specimen provides a good view of the spiracles, small tissue-covered openings usually located behind a ray's eyes. Rays use these openings to pump oxygen-rich water over their gills; this feature allows them to lie on the seafloor for long periods even though their gills are located on the bottom of their bodies.*

Marine Mammals

MARINE MAMMALS INCLUDE whales, dolphins, seals, sea lions, walruses, sea otters, and manatees. These species evolved from land mammals, and they have retained most familiar mammalian traits. They are air-breathing, warm-blooded vertebrates that bear live young, nurse their offspring, and have hair or fur at some point in their development.

The ocean presents some unique survival problems for marine mammals. Water absorbs body heat more rapidly than air does, and mammals must struggle to maintain their body temperature. Some solve this problem with blubber, thick insulating fat that keeps the cold out and the heat in. The fat may also provide energy reserves when food is scarce. Seals, sea otters, and sea lions have thick fur to help keep themselves warm.

PINNIPEDS

Walruses, seals, sea lions, and fur seals are all pinnipeds. The term comes from Latin words meaning fin footed. As divers, pinnipeds benefit from many special adaptations. The limbs of these animals have evolved into flippers. Their streamlined shape reduces water friction and makes swimming easier. Pinnipeds spend most of their time in the sea, but unlike dolphins and whales, they also have strong ties to land. They often come ashore to bask in the sun and rest, and they also gather on land or ice to breed and bear their young.

The five species of sea lions and nine species of fur seals form a group known as the eared seals. The eared seals are distinguished from true seals primarily by their obvious external ear flaps and by the structure of their hindlimbs, which they can rotate forward under their bodies. On land, the eared seals rely on their limbs for movement, while the true seals wriggle on their bellies and chests. In the water, the eared seals use their forelimbs in a breaststroke motion for swimming, and the true seals undulate their hindquarters for propulsion and use the forelimbs primarily for steering. The walruses form a third pinniped group that shares many of the physical and behavioral characteristics of the other two groups.

SEA OTTERS

Members of the weasel family, sea otters thrive along North America's western coast, where their playful antics keep observers entertained for hours. Their reddish brown to black fur is dense, soft, and very fine. Sea otters will venture onto land where they can travel

Opposite page: *A group of sea lions exchange greetings on tne beach at Seal Bay, Australia.* This page: *A sea otter feeds on an abalone in coastal Pacific waters.*

several hundred yards from shore, but they spend most of their lives at sea. Sea otters are the only marine mammals that lack blubber to keep them warm. They live in cold water, and like all other active mammals, they must maintain their body temperature. They solve this problem by consuming enormous quantities of food.

Otters are excellent swimmers and divers. They prefer to hunt in shallow water, to depths of 60 or 70 feet, but they sometimes venture deeper. Their hunting dives commonly last a minute or more, but they can stay submerged for as long as four minutes.

MANATEES

Manatees live in shallow, coastal waters in tropical regions; they never venture onto land and only rarely enter the deep ocean. Commonly called sea cows, manatees belong to the order Sirena, the sirens, because legend says that ancient sailors mistook them for mermaids.

The order includes four separate species. Three are manatees that live in the Atlantic basin—the West Indian, African, and Amazon manatees. The fourth is the dugong, which lives in the Indo-Pacific. In North America, manatees inhabit several Florida river systems, and they thrive in the coastal estuaries of many Central American countries, especially Belize.

CETACEANS

The order Cetacea consists of dolphins and whales. Collectively they are the oldest and best-adapted group of aquatic mammals. Their forelimbs have evolved into flippers joined to the body at the shoulder, and they have completely lost their hindlimbs. Their large tail flukes and dorsal fin greatly assist them in their aquatic lifestyle, as does the blowhole atop their head, which serves as an airway to the lungs.

There are two major categories or suborders of cetaceans. Baleen or filter-feeding whales are in the suborder Mysticeti, and toothed whales are in the suborder

Above: *A California gray whale pokes its barnacle-covered head through the surface for a breath of air.*

Odontoceti. The two groups differ in several major ways, but the biggest difference is that the baleen whales lack teeth after birth. They have instead long horny plates called baleen that look somewhat like the fibers of an oversized toothbrush. The baleen filters great quantities of plankton, such as krill, copepods, and isopods, and some small fish from the water when baleen whales feed. All 11 known species of filter-feeding whales use their baleen as the primary way to catch plankton and small fish.

Baleen whales usually migrate annually from their polar or subpolar feeding grounds to their calving and breeding areas in temperate and tropical seas. These whales usually spend the warmer months in their feeding areas before moving toward the equator during polar winters.

Because plankton occurs densely in shallow waters, baleen whales spend much of their lives in the top 150 feet of the water. However, they are very capable swimmers, and their dives often last as long as 20 minutes.

The toothed whale group includes a number of species, such as killer whales, beluga whales, narwhals, and pilot whales. It also includes those species commonly called dolphins and porpoises. While this surprises many people, it is taxonomically correct. The whales are generally larger than the dolphins and porpoises, but these animals still share many traits.

Having teeth enables this group to pursue individual prey rather than filter feeding like the baleen whales. This means that they can populate a wider range of areas where plankton is not abundant.

Above: *Ivory-colored beluga whales* (Delphinapterus leucas) *inhabit the icy waters of the northern polar region.*

Over time, toothed whales have diverged and specialized. Several species live exclusively in freshwater, others are strictly marine, and still other species can live in freshwater or salt water. This diversity also means that several species can live near one another. Different species tend to pursue different foods and therefore occupy slightly different niches, so they can inhabit the same region without competing too heavily for food.

🐟 Opposite page: *Galápagos sea lions live exclusively in the Galapagos Islands, where they spend most of their time swimming in the cool waters of the Humbolt Current. Males often grow to over 700 pounds, while females typically reach only about a third of that weight. The Galápagos sea lion is a close relative of the well-known California sea lion; the principal difference is the slightly smaller, more slender skull of the Galápagos variety.*

🐟 Above: *A California sea lion mother tends to its newborn pup. As with most mammals, the sea lion pups receive a great deal of parental care; pups remain in close company with their mothers for nearly their first whole year of life. Like most pinnipeds, California sea lions are very social animals. Large groups of them are quite common in coastal areas ranging from British Columbia to the Galápagos Islands near Ecuador.*

🐋 Above: *A Steller sea lion hauls out on a rock and emits a characteristic bark. Restricted to the Northern Hemisphere, Steller sea lions range across the North Pacific Rim from Japan to California. Males can reach over 1,000 pounds, easily making them the largest of the sea lions. They are named for a German surgeon and naturalist who described the animals while shipwrecked on Bering Island in 1742.*

🐋 Right: *Gregarious Steller sea lions gather in the chilly waters off Alaska. Here, walleye pollock are likely to make up an important part of the sea lions' diets; in Canadian waters, Steller sea lions mainly eat herring, rockfish, cod, squid, and octopus. Some will travel up freshwater rivers, especially in Oregon, where they prey primarily on lampreys and salmon. Young male Steller sea lions will also attack and eat the pups of certain seal and fur seal species, although this is by no means a staple of their diet.*

🐟 Opposite page: *The creamy markings and graceful antics of Australian, or blonde, sea lions make them among the most stunning of the pinnipeds. Nonmigratory, they occur only in South Australia. Like many pinnipeds, they are a favorite food source for great white sharks. They also face human influences that make them perhaps the most severely endangered pinnipeds. South American sea lions and Hooker's sea lions of New Zealand are the other Southern Hemisphere species.*

🐟 Above: *These maturing male Australian sea lions may well be competing for territory—and the right to mate with the females in it. Normally, a territory will contain four to six females, referred to as a harem. Males behave aggressively toward each other and ruthlessly herd the females. For breeding, Australian sea lions favor the sandy beaches and smooth rocks of the coast. At birth, pups are chocolate brown; they will be a gingery color when they begin molting at about two months.*

🐋 Opposite page: *A northern fur seal bull, easily recognized by his great size and brownish-black fur, sits with his harem on Alaska's Pribilof Islands. During the spring breeding season, tens of thousands of pinnipeds converge on this island rookery. In nearly all pinnipeds, males and females gather for birthing and mating at the same time. Females often breed only days after bearing their young. In most species, the resulting fertilized eggs do not settle in the uterine wall to develop until several months later. This delayed implantation allows a new mother to nurse her pups and regain her strength unhindered by her next pregnancy.*

🐋 Right: *A harbor seal pokes its head out of northern polar waters. Harbor seals have a chunky shape and lightly colored coat covered with dark spots, and they often show considerable curiosity toward snorkelers and divers. One of the smaller Arctic species, they actually range throughout the temperate and polar waters of the North. The other seals of the Arctic Circle include ringed seals, hooded seals, bearded seals, and harp seals.*

Above: *A harp seal pup's snowy white coat shimmers in the sunlight. Harp seals have achieved worldwide attention recently due to the heavy pressure and needlessly brutal harvesting methods of hunters in Greenland and Newfoundland who value the pups' white fur. New restrictions limit the hunting of newborns, but much controversy remains over their future and that of their parents.*

Left: *A harp seal emerges from a hole in the ice in the Gulf of St. Lawrence. Air holes such as this one provide haven from predators as well as access to food and air; they are often a hotly contested commodity. Seals will sometimes fight over the rights to use a specific air hole, or an individual may claim a particular hole as part of its territory and guard it aggressively against any encroachers.*

🐋 Opposite page: *An aggressive encounter between two female harp seals* (Phoca groenlandica). *The common name comes from the distinctive harp-shaped or lyre-shaped pattern on their backs that becomes prominent at maturity. Migratory animals, harp seals range far into the Arctic in the summer to hunt herring, cod, capelin, squid, and crustaceans. As winter and harsh weather approach, they travel southward to the pack ice on the coastal regions of the North Atlantic.*

🐋 Above: *A harp seal pup suckles its mother. Births generally occur from late February through mid-March. Weighing approximately 25 pounds at birth, pups will weigh in at 80 pounds when weaned about two weeks later. During the brief nursing period, the mother fasts or eats very little; she will lose approximately seven pounds of body mass each day, mostly in the form of blubber. Mother harp seals will defend their pups if approached by other seals.*

Above: *Two northern elephant seal males joust on San Benitos Island during the breeding season. They probably won't mate this year because older, larger bulls will drive them from the beach. These two young bulls lack the large, bulbous nose characteristic of a mature male (left); the nose becomes prominent later in life and serves as a display during competition. When a mature male is able to win a breeding territory, he must constantly defend it from other males in grueling, sometimes brutal battles. Competition is so severe that frequently males will die only a few years after maturity, debilitated by the rigors of mating and defending.*

 Left: *Southern elephant seals, which live in the Indian, South Atlantic, and Antarctic oceans, are the largest seals in the world. Full-grown males can be as long as 15 feet and weigh as much as 5,000 pounds; females are considerably smaller, ranging from a third to a half the size of the males. For females, maximum longevity is about 23 years; for males, it's about 20 years, but few ever reach that age. Cephalopods and fish are the primary diet for both males and females.*

Opposite page: *Most pinnipeds feed largely upon fish, mollusks, and crustaceans, but the leopard seal is a notable exception. Prowling the waters of the Antarctic, it also feeds on penguins and other seals. A favorite trick is to lurk under holes in the ice floes and suddenly rise up and grab any victim that has ventured too near the edge of the ice. The leopard seal's varied diet affords it a broad distribution.*

Above: *A mother Weddell seal tends to her pup. One of only four seal species to make their home in the Antarctic, Weddell seals live farther south than any other mammal. During the harsh winters, they spend much of their time in the water, which offers a warmer, more hospitable environment than the frozen, storm-swept ice cap. Using specially modified canine teeth, they bore breathing holes through the ice and constantly scrape away at them to keep them from freezing over.*

🐋 Above: *Walruses are famous for their impressive tusks, which are actually canine teeth that continue to grow throughout the animals' lives. The tusks of females are short and curved, while in males they are longer and straighter. Older males can have tusks exceeding three feet and weighing up to 12 pounds each. Walruses use their tusks to pull and brace themselves as they travel the icy Arctic landscape; mature males also use them to battle for territory and to defend against predators.*

🐋 Opposite page: *On Round Island in Alaska, walruses sunbathe and sleep on the rocky shore. Herds may consist of a thousand or more animals. Found only in the Arctic, walruses are among the largest pinnipeds in this region. Adult males reach sizes of 11 feet and 3,500 pounds. Walruses almost never stray far from coastal areas; they rummage through the seafloor in these shallow areas, hunting primarily for clams and other shellfish.*

🐋 Above: *Walruses are highly gregarious animals, and they commonly gather in groups of hundreds. Like other pinnipeds, they follow very specific mating patterns. They breed every other year, and 15 months later the female bears a single pup. A newborn walrus is about four feet long and weighs close to 110 pounds. The female nurses her young for up to two years.*

🐋 Opposite page: *Pinnipeds have had to develop some unusual and effective strategies to regulate their body temperature. Dense fur and a thick layer of blubber provide insulation against the chilly climate and icy waters of the poles. As another way to conserve heat, many pinnipeds can stop the flow of blood to the surface of the skin by constricting their blood vessels. When they do so, they may appear almost white, as this walrus demonstrates.*

Above: *Sea otters are one of only a few types of animals definitely known to use tools. They frequently lie on their backs at the surface of the water and crack open clams, mussels, and other hard-shelled prey with rocks. Below the water, they wield rocks like hammers to knock scallops and mussels off the substrate. Known for their voracious appetites, mature male otters can consume 5,000 pounds of food each year.*

Opposite page: *A sea otter frolics in the waters off Monterey Bay. Sea otters once ranged from the shores of Mexico's Baja peninsula north to Alaska, eastward across the Bering Sea, and south to Japan. However, after 170 years of being intensely hunted for their pelts, the otters' numbers and range have been greatly reduced. Today they are making a comeback, but their success is fraught with controversy. Fisheries that depend for their livelihoods on the same shellfish that make up the otters' diet fear competition from a large otter population.*

Left: *A Florida manatee swims in the Crystal River. Adults consume up to 100 pounds of plant life a day to maintain their weight, which averages 1,000 pounds but can exceed 3,300 pounds. In the winter, Florida manatees swim up to freshwater springs where the water is warmer than in the ocean. There, they perform a service to humans and other animals by feasting on the prolific hyacinth that would otherwise choke the inland waterways.*

Opposite page: *The Indo-Pacific dugong has rubbery lips covered with bristles, as does its close relative the manatee. As herbivores, dugongs and manatees have strong molars for grinding and chewing, but they lack cutting teeth. They use their flexible flippers for paddling through the water and for balancing on the seafloor. Slow-moving and virtually defenseless, they are probably easy prey for sharks, alligators, and crocodiles, and many are killed by hunters and careless boaters.*

🐋 Above: *A snowy white beluga whale surfaces off the Siberian coast. Beluga whales inhabit shallow waters. They often swim in estuaries and bays, and during summer months they may travel more than a hundred miles up some rivers. At birth, their bodies are dark brown with bluish-gray spots; within a short time, the spots fade and belugas take on a stunning ivory color. As adults, they are pure white.*

🐋 Opposite page: *Like all whales, belugas are lithe and graceful despite their great size. Residents of the far northern Pacific and the Gulf of St. Lawrence, beluga whales are not well studied in the wild because they rarely venture into more hospitable waters. The bulge, or melon, on the top of a beluga's head is filled with fatty tissue saturated with oil. Researchers believe the melon may be a part of this toothed whale's echolocation system.*

🐋 Left: *A female narwhal swims with her pup. Another interesting species of toothed whale, the narwhal's most notable feature is a highly developed tooth, or spiral tusk, that extends out from the male's upper jaw. Adult males reach a length of about 14 feet; the tusk adds another six to eight feet to their length. Narwhals are an Arctic species. They feed on squid, fish, and a variety of crustaceans.*

🐋 Right: *Reaching a size of 60 feet and 45 tons, sperm whales are the largest toothed whales. Their name comes from the spermaceti oil that made them a favorite target of whalers. Sperm whales feed on a variety of foods, but deep-sea squid are a major part of their diet. The whales dive as far as 2,500 feet, perhaps even deeper, as they search for food. The longest recorded dive of a sperm whale lasted two hours and 18 minutes.*

🐋 Above: *This jet-black pilot whale shows off its large, rounded head, somewhat defined lips, and conical teeth. Less evident is the white anchor-shaped patch that may extend from the whale's chin down along its underside. Pilot whales were named by European fishermen who followed the animals, believing that schools of herring could reliably be found swimming beneath them.* Right: *The two distinct species of pilot whales are set apart primarily by the length of their flippers. The short-finned variety tends to favor the warmer areas of the temperate seas, while the long-finned group often ventures into more northern areas. They are both deep-diving species that feed primarily on squid.*

🐋 Above: *A pair of killer whales leap and roll across the surface of the water. Breaching like this is a common natural behavior in whales and dolphins. The activity may help to rid the animals of parasites, or it may be a way for them to navigate by looking for familiar coastal landmarks. They might also be doing it simply for the joy of it.*

🐋 Left: *Killer whales are found in virtually all oceans, although they are most common in the cooler areas of the Pacific and Atlantic. Like most toothed whales, they are social animals. They often live in pods of up to 30 individuals. Their group behavior may be matriarchal, meaning that a female assumes leadership in each pod. Different pods will sometimes exhibit different behavior patterns. Some groups are wanderers, while others tend to stay in a particular area. One group might focus its hunting efforts on pinnipeds, and another might feed primarily on large fish.*

Right: *The well-developed teeth of a killer whale allow it to pursue fish, squid, penguins, seals, sea lions, and even other whales. Like many other toothed whales, they often hunt in organized packs in order to corral their prey. In arctic regions, killer whales will repeatedly strike ice floes from below, sometimes shattering them, to drive seals and other prey from the safety of the ice into the water. They sometimes toss captured prey about in the water for several minutes before eating it. While this may seem cruel, the practice probably serves to sharpen their hunting skills.*

Opposite page: *A killer whale spy hops, observing the surface of the Pacific Ocean off British Columbia and drawing a breath of air. Like other species of toothed whales, killer whales use a form of natural sonar called echolocation to hunt and to learn about their surroundings. They emit sound waves that strike objects in the water and reflect back to them. The whales analyze the reflected waves to identify objects and gauge their distance.*

🐋 Above: *People know many species of toothed whales as dolphins and porpoises. They are typically smaller than the species called whales and there are some physiological differences, but all toothed whales are quite closely related. In fact, some species we commonly refer to as whales actually belong to the dolphin family. Taxonomists distinguish among the groups of toothed whales primarily by the structure and arrangement of the teeth. The five families of marine toothed whales are the sperm whales; the beaked and bottlenose whales; the oceanic dolphins; the porpoises; and the belugas and narwhals.*

🐋 Opposite page: *The common dolphin has a long, slender body and well-defined features. It is one of only a few dolphin species to carry markings of three colors—white, black, and cream. The crisscross pattern generally found on their flanks is another identifying feature. Agile swimmers, many dolphin species can maintain speeds of 20 miles an hour by altering their skin's shape as they swim to lessen water resistance. The dolphins' highly social lifestyle, complex playing behavior, and curiosity toward humans give strong support to the generally held notion that dolphins are highly intelligent animals.*

Opposite page: *A spotted dolphin cow with calves. Like most dolphins, spotted dolphins aggregate in pods. Pods containing several hundred Atlantic spotted dolphins are not unusual. Their counterparts from the Pacific Ocean sometimes form groups that contain between 3,000 and 4,000 individuals. They will also freely associate with other species of dolphins. Friendly, inquisitive animals, spotted dolphins will swim within touching distance of divers and swimmers.*

Above: *A spotted dolphin cow swims with its calf. Calves lack the characteristic spots for which the species is named; these will develop as the animal matures. Calves are thought to be born after a gestation period of perhaps 11 months. Research suggests that female spotted dolphins give birth every two years; stress placed on the dolphins due to tuna fishing and resultant higher mortality rates may have worked to accelerate this process.*

🐋 Opposite page: *Right whales got their name because they were the best, or the right, whale for whalers to hunt. These baleen whales travel slowly, are relatively easy to find and kill, float when they are dead, and have great commercial value. Sadly, their fate remains in question today due to the intense hunting pressure exerted in the last century. An estimated stock of 100,000 animals only a few hundred years ago has dwindled to a population of no more than 4,000.*

🐋 Above: *A southern right whale calf swims safely and securely beneath its mother's belly. Mothers and babies maintain a close relationship. Typically, newborns remain with their mothers for an entire year. As this calf matures, the hard, whitish patches visible on the tip of its jaw and around its eyes will become more pronounced. These callosites are unique from animal to animal, and whale researchers can use the callosites as a way to identify individuals.*

🐋 Above: *Every year, huge numbers of gray whales make the long passage from summer feeding grounds in the Bering Sea off Alaska to calving grounds along the Baja peninsula in Mexico. The 10,000-mile round trip is the longest mammal migration known to science. The journey allows the grays to avoid the harsh weather conditions and limited food supply of the Arctic winter. The warm, protected lagoons of their winter home may also afford the new offspring a better chance of survival.*

🐋 Opposite page: *Unlike most other large whales, gray whales usually favor shallow water over deep. They will sometimes allow boaters and divers to touch and pet them, which is why they are often known as friendlies. Gray whales are thought to live 50 years or more. Their dorsal region will typically have a distinctive patchy look to it. This pattern is the result of the area being infested by barnacles and by crustaceans called whale lice.*

🐋 Above: *A blue whale shows off its tail, or flukes. Whale flukes are as distinctive as an individual name tag. Scientists often use the unique flukes to catalog wild whales and track their movements. A large blue whale's flukes can stretch 15 feet from tip to tip. Its mouth can be 20 feet long, its heart can be the size of a small car, and its stomach can be large enough to hold over two tons of food.*

🐋 Right: *A blue whale exhales. A whale's distinctive spout comes from the moisture in its warm breath and the seawater that pools over its blowhole. The blue whale's blow produces a 30-foot-tall spout. The baleen or filter-feeding whales are often referred to as the great whales due to the enormous size of many species. Not only is the blue whale the largest filter-feeding whale, it is also the largest animal living on Earth today. Blue whales can grow to an incredible 110 feet and weigh as much as 120 tons.*

🐋 Above: *This humpback whale clearly shows the throat grooves or pleats that distinguish the six species of rorqual or finback whales. The pleats run lengthwise down the throat to the navel area. Working like the folds of an accordion, they expand the throat greatly when the whales feed, allowing them to take in huge quantities of water, which they strain through their baleen for plankton and small fish. Humpbacks generally have between 14 and 24 throat grooves; other species of rorqual whales may have as many as 100.*

🐋 Opposite page: *The humpback has several features that distinguish it from the other rorqual whales. Its body is stout and rounded, while the other five rorquals (blue, fin, sei, Bryde's, and minke whales) tend to have a more slender shape. The humpback's tail fluke is distinctive also; the edges are sharply serrated, and it has noticeable points on the ends and a dramatic notch in the center. The animal's front end is also unique, with a series of pronounced bumps containing hair follicles on the top of the head and around the mouth.*

🐋 Above: *In an awesome display, a pair of humpback whales breach in the waters off Hawaii. The flippers of a humpback can be a quarter to a third the length of the entire animal. The long, flexible flippers contribute to the whale's exceptional maneuverability below the surface. The unusual knobs on the forward edge of the flippers are an indication of the whale's prominent finger bones.*

🐋 Right: *All 11 known species of filter-feeding whales use their baleen as the primary way to catch food, but some species also use other techniques. Humpback whales like these have devised a remarkable system called bubble netting for feeding on small schooling fish. A humpback will swim in a spiral below the school and slowly exhale to create a circular curtain of bubbles that herds the fish into a tighter and tighter school. With mouth agape, the whale then rushes straight upward through the frightened, confined group and captures as many fish as it can.*

 Above: *Humpback whales are well known for their melodic songs that consist of a complex series of moans, roars, and high-pitched chirps, but many questions remain unanswered about this behavior. Males seem to be the only ones to sing, and they usually do so only in the warm-water breeding areas. They could be defining territory, threatening other males, or attempting to attract a mate. They typically sing in groups, though, with all the members creating the exact same series of sounds for 20 minutes or more and then repeating that same song or a slight variation of it.*